Prisms
in the Medical and Surgical Management of Strabismus

PRISMS
in the
Medical and Surgical
Management
of
STRABISMUS

Suzanne Véronneau-Troutman, M.D. FRCS(C) FACS

Clinical Associate Professor,
Department of Ophthalmology Cornell Medical College;
Associate Attending Surgeon, The New York Hospital–Cornell Medical
Center, Chief Motility Clinic;
Attending Surgeon, Manhattan Eye Ear and Throat Hospital,
Chief Motility Clinic,
New York, New York

 Mosby

St. Louis   Baltimore   Berlin   Boston   Carlsbad   Chicago   London   Madrid
Naples   New York   Philadelphia   Sydney   Tokyo   Toronto

Mosby
Dedicated to Publishing Excellence

Editor: Laurel Craven
Associate Developmental Editor: Wendy Buckwalter
Project Manager: Gayle May Morris
Production Editor: Dana Peick
Manufacturing Supervisor: Kathy Grone

Printed in the United States of America
Composition by The Clarinda Company
Printing/binding by The Maple-Vail Book Manufacturing Group

Mosby–Year Book, Inc.
11830 Westline Industrial Drive
St. Louis, Missouri 63146

International Standard Book Number: 0-8016-5819-5

94 95 96 97 98 / 9 8 7 6 5 4 3 2 1

Preface

My approach to ocular motility disturbance has changed since the 1969 publication of the Hugonniers' book, which I translated into English and edited. It is because of the poor results of pleoptics and other instrumental methods for the treatment of deeply rooted sensory anomalies that I became interested in prismotherapy as advocated by Pigassou-Albouy and Bérard.

From 1972 to 1978 I conducted a course at the American Academy of Ophthalmology and Otolaryngology, which I continued until 1987 under the auspices of the newly formed American Academy of Ophthalmology. The course was first entitled "Recent Advances in Strabismus Therapy—Fresnel Prism Membrane" and later became "Prisms in the Surgical and Medical Management of Strabismus." Although I was asked to continue the course in 1988, I decided I could better cover the topic in a textbook. During the same period, among all the papers I presented in the United States and abroad that related to strabismus, more than 30 directly addressed the use of prisms.

In 1978 my thesis for the American Ophthalmological Society was "Fresnel Prisms and their Effects on Visual Acuity and Binocularity." It was this study that encouraged me to confront the limitations of the Fresnel prism membrane and the possibilities of the Fresnel principle as it applied to solid rather than flexible plastic material.

In 1980 in an interview with Dr. Benjamin Boyd for the Highlights of Ophthalmology, Volume I (Silver Anniversary Edition 1981, pages 482-483), I cited the three most important advances in strabismus of the 1970's: (1) the Fresnel prism membrane, (2) the reintroduction of adjustable sutures, and (3) the availability of finer suture materials, better surgical needles, and the use of the surgical microscope. This last advance has only just begun to be endorsed by other strabismologists, and the efficacy of the first has continued to be reevaluated throughout the intervening years.

It is common to hear from European colleagues that the American school of strabismus is too "motor oriented," that we are mainly interested in putting the eyes straight, and that we do not take the time to cure sensory anomalies. Having been exposed to both schools, I believe that nonsurgical methods continue to receive our careful scrutiny.

The verbal and written communications from colleagues who have attended my courses and lectures and have visited me in my office, and communications from the residents, fellows, and orthoptists who have worked with me have all indicated the desire for a comprehensive textbook on the use of prisms in the medical and surgical management of ocular motility disturbances. This volume attempts to address this need and do so in a simple, practical way. With few exceptions, the description of tests is limited to those associated with prism use. Surprisingly they cover most of the tests used in the office setting for the sensory and motor evaluation of ocular motility disturbances.

As stated in the introduction, the purpose of this book is to be practical, but it does not prevent opening a door to the future. It is for this reason that topics such as laser-induced corneal prisms and binary optics are addressed with the hope that future exploration and application will be encouraged.

Acknowledgments

I would like to thank the following:

Dr. Richard Troutman, my husband, who has encouraged me throughout the writing of this book and who is responsible for all the computer-designed illustrations. He has been an invaluable help in the final manuscript preparation. His interest in a different field of ophthalmology has allowed him to remain an objective critic.

Ms. Eileen Traykovski CO who has been working with me for 15 years. She collected and summarized the data of the case presentations that had been examined by both of us in detail. She also provided good suggestions and personal support.

Ms. Joan Baras CO who has been indirectly involved but who, nonetheless, deserves my thanks.

I would like to thank Dr. Eugene Helveston, a good friend and trusted colleague, for writing the Foreword for this book.

Suzanne Véronneau-Troutman

Foreword

The diagnosis and treatment of eye disease by the generalist and the subspecialist employ the ophthalmic prism or the effect of the ophthalmic prism in nearly every patient encounter. The applications of prisms range from intensive and long-term therapy for strabismus to accessory use in ophthalmic instruments. Although omnipresent in ophthalmic practice, prisms are acknowledged primarily as a unit of measure, (e.g., "the patient has 20 prism diopters of esotropia.") Until now, the ophthalmic prism has not been the subject of a scholarly, comprehensive, and usable book. Finally, Doctor Veronneau-Troutman has provided a thoroughly competent book that offers a lucid discussion of the opthalmic prisms beginning with theory and ending with a meticulously detailed description of their myriad clinical uses. This book fills a void that has existed far too long.

The physics of ophthalmic prisms can be daunting, but with this book, an overall understanding of the concepts is within the grasp of those who use them. These principles are presented clearly, establishing a sound basis for understanding the effect of prisms in diagnosis, instrumentation, and therapy. Prism therapy has been largely neglected in current textbooks that stress optics and surgery.

With extensive experience in the therapeutic use of prisms, the author has provided a book that is both easy to read and informative. Her enthusiasm for all aspects of prism use is infectious and engaging. Whereas the advent of Fresnel press-on prisms has resulted in an increased use of prisms, the author's involvement clearly predates this trend. Her interest in prisms is not a response to a fad, but rather it is a manifestation of a longstanding, sincere commitment.

It takes a unique person with genuine dedication to undertake the challenge that the subject matter of this book presents. The author has accepted this challenge, approached the task forthrightly, and has succeeded in producing a wonderful book on a subject that has certainly up to now deserved more attention than it has received.

Eugene M. Helveston, MD

Contents

Prisms
in the Medical and Surgical Management of Strabismus

Introduction

Conventional prisms have been used for more than a century[1-7] to evaluate the motor and sensory condition of ocular motility disturbances. The principles that govern their use and their limitations have been addressed from time to time and should be familiar to all those who use prisms.

The introduction of Fresnel wafer prisms in 1965 and of Fresnel membranes in 1970 has revived interest in prisms as a diagnostic tool as well as a therapeutic tool. The prism adaptation test (PAT) that evolved from the availability of these thin, light-weight, high-power prisms is rarely done as originally described. However, the test did motivate an in-depth prospective multicenter study, supported by the National Institute of Health, on the efficacy of PAT in the surgical management of acquired esotropia. Among my many indications for presurgical prisms, I have found PAT to be especially useful for esotropic patients with normal retinal correspondence (NRC), as well as for patients with intermittent exotropia and patients with A and V patterns. The application and interpretation of the test is discussed for each type of motility disturbance.

The PAT must be differentiated from prismotherapy, which necessitates a different approach. Prismotherapy has to be carried out for months, whereas a PAT can be done within 2 hours to a maximum of 2 weeks. The technique also differs. For example, in a patient with NRC under prismotherapy, the minimum power that controls the deviation is prescribed. The stronger prism is placed over the dominant eye to overcome suppression. When prismotherapy is completed and if surgery is indicated, PAT is performed to elicit the full angle of deviation. For PAT, in contradistinction to prismotherapy, equal prisms are applied, or a weaker prism is applied over the dominant eye, to avoid changing the eye preference.

As an examination tool the Fresnel Prism Trial Set, using hard plastic prisms (12Δ to 40Δ), has definite advantages over Fresnel membranes and conventional prisms of corresponding powers. In my opinion the Fresnel Trial Set is an essential part of standard strabismus instrumentation. *To make the best use of prisms, the examiner has to use them routinely and also be selective in their application.*

When referring to a previously advocated treatment, it is important also

1

to research the most recent publication or current practice of that author, which may differ from his or her earlier view. Insofar as possible, such an approach is followed throughout this book.

The old observation that constant esotropia of very early onset has a poor sensory prognosis has not been substantially changed either by early surgery (between 6 and 12 months of age) or by early or delayed prismotherapy. Nevertheless, should parallelism be restored even earlier, in the first few months of life, the results achieved up to the present time may change. Acquired esotropia has a better prognosis, but if prismotherapy is used, strict guidelines must be followed. In my experience, prisms are more effective in curing a binocular defect in primary exotropia than in esotropia; for this former group of patients, prismotherapy is particularly indicated.

Prisms to compensate diplopia are used in a much greater population of patients and are useful with concomitant as well as with incomitant deviations. Unfortunately, many patients with incomitant deviations are often denied a prismatic correction although they can usually benefit from it. Finding the correcting prism, often an oblique one, can be time consuming. However, with practice, taking advantage of simple and efficient methods, an accurate prismatic correction can be prescribed after only a few minutes. It is rewarding to successfully correct a diplopia in a patient whose problem has been ignored or managed unsuccessfully up to that time.

Although the purpose of this book is to present the practical use of prisms in the management of ocular motility disturbances, it does not prevent opening a door to the future. With broader interest and a corresponding increase in demand, we should see an improvement in the quality and optics of ophthalmic prisms. With better tools, conditions that until now have not responded to prismotherapy will require reevaluation. Laser-induced corneal prism may not be far away from its clinical application. This new modality is discussed in Chapter 1. Binary optics, an advanced implementation of the Fresnel principle, is discussed in Chapter 2. This technology only recently has become commercially feasible and may provide the ideal ophthalmic prism.

REFERENCES

1. Graefe AC: *Klinische analyse der motilitätsstörungen des auges,* Berlin, 1858, Peters.
2. Hart E: A mirror of the practice of medicine and surgery in the hospitals of London, St Mary's Hospital (Ophthalmic Department), case of extreme squint cured without operation by the use of prisms, with clinical remarks, *Lancet* 119, 1864.
3. Javal E: *Manuel théorique et pratique du strabisme,* Paris, 1896, Masson.
4. Krecke FWC: Brillen voor scheelzienden, *Ned Lancet* 3:1047, 1847.
5. Maddox E: *Clinical use of prisms and decentering of lenses,* Bristol, Eng, 1889, J Wright.
6. Maddox E: *Clinical use of prisms,* ed 5, Bristol, Eng, 1907, J Wright.
7. McFarland SF: A personal experience with prismatic glasses, *Trans Am Ophthalmol Soc* (19th meeting) pp 479-481, 1883.

CHAPTER 1

Optical Principles of Prisms

An ophthalmic prism is a transparent triangular wedge of refracting material. The thicker portion of the wedge is the *base,* and the thinner portion opposite is the *apex.* The angle between the polished nonparallel opposing surfaces (faces) of the prism, which meet to form the apex of the prism, is termed the refracting or *apical angle.* The line bisecting this angle is called the *axis* of the prism. The straight line at the apex at which the two faces of the prism meet is the *edge.* The *principal section* of a prism is a section made by a plane perpendicular to the edge of the prism. Ophthalmic optics is concerned only with rays of light traveling in principal sections (Fig. 1-1).

The angles of incidence and emergence relate to the rays striking the first plane surface and emerging from the second. The total deviation is the net change in direction of a ray produced after refraction by both surfaces of a prism. Although the deviation is toward the base of the prism, to the observer the object appears displaced toward the apex (Fig. 1-2). *The total deviation equals the sum of the angles of incidence and emergence minus the apical angle of the prism.* This formula derives from the following trigonometric calculations (Fig. 1-3).[6]

A ray, DEFG, passes through a prism, ABC. The refracting or apical angle is represented by a. The angles of incidence are formed at the two faces by i and i', and the angles of emergence are formed by e and e'. We can see that the angle

$$AEF = 90° - e' \text{ and } AFE = 90° - i'$$

Hence,

$$a + (90° - e') + (90° - i') = 180°$$

$$a = i' + e'.$$

The deviation at AB is measured by i − e' and at AC by e − i'.

$$D = i - e' + e - i'$$

$$= i + e - (e' + i')$$

$$D = i + e - a$$

3

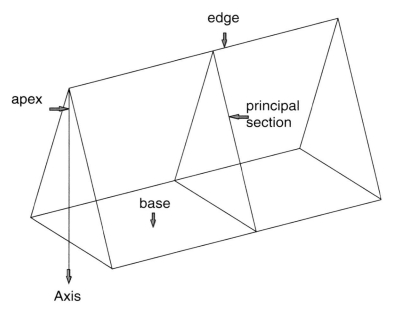

FIGURE 1-1. The ophthalmic prism.

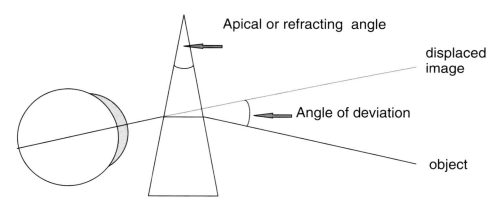

FIGURE 1-2. Prism displacement. Image of object appears displaced toward the apex of the prism.

The refracting angle or apex of a prism is always specified in degrees (°). The angle of deviation or refracting power of an ophthalmic prism is always specified in prism diopters (Δ).

One prism diopter produces an apparent linear displacement of 1 cm of an object situated 1 m away. 1Δ = 1 centrad = 0.57°.

The prism diopter represents a tangent measurement, where the degree and the centrad are units of arc measurement. As prism powers increase, these reciprocal values are not linear and lose accuracy. For the prism powers used in ophthalmology, each degree of angular deviation closely corresponds to 2Δ. Although the centrad may be more accurate for high-

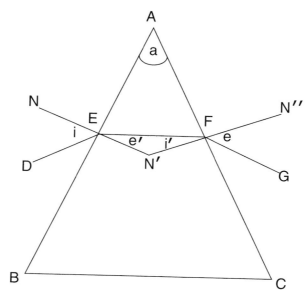

FIGURE 1-3. Refraction by an ophthalmic prism.

power prisms, the prism diopter unit promoted by Prentice[15] has gained general acceptance.

It is important not to confuse the relationship between degrees and prism diopters and the relationship between the refracting angle and angle of deviation. For example, a prism made of crown glass with an index of refraction (n) of approximately 1.5 and a refracting angle of 1° induces 1Δ of deviation. Therefore, for the comparatively thin prisms used in ophthalmology, *the angle of deviation (D) is equal to half the refracting angle (apical angle; a).* When the angle of incidence at the first interface is equal to the angle of emergence at the second interface, the total deviation is the smallest possible, the *minimum deviation* for that particular prism. In ophthalmology the prisms used are comparatively thin, and in the majority of cases the rays pass through symmetrically.

Therefore the formula $D = i + e - a$ can be simplified to

$$D = 2i - a$$
$$i = \frac{D + a}{2}$$

According to Snell's law: $n \sin i = n' \sin r$

If both angles are small, their sines are equal to the angles.

$$\frac{D + a}{2} = \frac{na}{2}$$
$$D = a (n - 1)$$
$$D = a (1.5 - 1)$$
$$D = \frac{a}{2}$$
$$D = 0.5° = 1Δ$$

The *effective power* of a prism is influenced by the way a prism is held because this determines the angle of incidence.

A *total internal reflection* takes place when a ray of light (NN′) strikes the hypotenuse of the prism at an angle of incidence greater than the critical angle. The light ray is then totally internally reflected and emerges as N′N″ deviated at a right angle from its initial direction (Fig. 1-4). When the index of refraction is known, the critical angle for a given prism can be found by applying Snell's law, or conversely the critical angle can be changed by modifying the index of refraction of the media involved. For a crown glass prism with an index of refraction of 1.523, the critical angle is found by the following equation:

$$n \sin i = n' \sin 90°$$

$$i \text{ critical} = \sin^{-1} \frac{nr}{ni}$$

$$i \text{ critical} = \sin^{-1} \frac{1.0}{1.523} = 41°$$

The reflecting prism in the eyepiece of an indirect ophthalmoscope provides increased light to the observer's eye by total internal reflection. A fiberoptic tubule conducts light by multiple total internal reflections.

Prentice position. *In the Prentice position*[15] *the total prismatic deviation occurs at only one of the two faces of the prism* (Fig. 1-5, A). At one face, the light ray strikes at 90° to the surface, therefore inducing no deviation. The deviation occurs only when the ray strikes the second nonparallel surface. For this

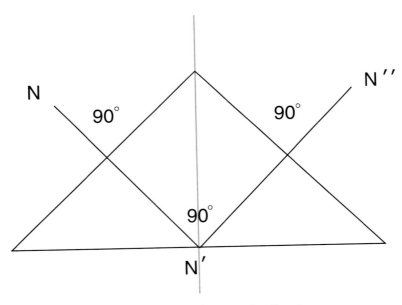

FIGURE 1-4. Total internal reflection.

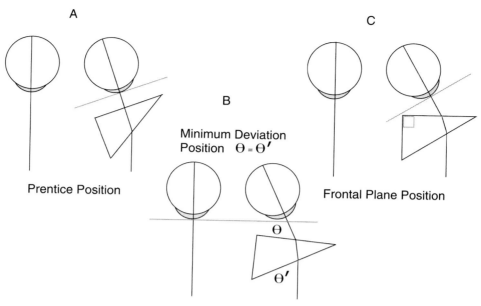

FIGURE 1-5. Positions for prism measurement. **A,** Prentice position. **B,** Minimum deviation position $\Theta = \Theta'$. **C,** Frontal plane position.

position, the prism is held with one face perpendicular to the line of sight or parallel to the iris plane. *A specific power prism held in the Prentice position necessarily has a power greater than if placed in the position of minimum deviation* (the angle of incidence equals the angle of emergence) (Fig. 1-5, B). The Prentice position power is the power normally specified for ophthalmic prisms, because it is more convenient clinically to mount a prism with one face perpendicular to the line of sight. *Trial prisms designed for a trial frame, Fresnel prisms, and incorporated prisms in spectacles are all calibrated for the Prentice position.* Their powers are measured in the same position with the rear face of the lens against the nose cone of the lensmeter. Loose, square glass prisms are also calibrated in the Prentice position.

Prentice (1890) was mainly concerned with the therapeutic prisms of low powers (up to 8Δ) used at the time. After Duane popularized the prism and cover test to measure strabismus, which made use of higher power prisms, it was Hardy in particular who pointed out the important measurement errors that could be induced by incorrect prism positioning.[10;11]

Minimum deviation and frontal plane position. The position of minimum deviation (Fig. 1-5, B) was advocated by Jackson many years ago.[12] In the United States the loose, square plastic prisms set and Beren's bar of prisms are calibrated in the position of minimum deviation. For distance measurements in primary position, they are held with the rear surface of the prism parallel to the frontal orbital plane (frontal plane position) (Fig. 1-5, C), because this most closely approximates the angle of minimum deviation. In this orientation, for measurements at distance in primary position, "the posterior face of the prism is perpendicular to the direction of the fixation object along the line of sight."[11;21] For other measurements in the diagnostic positions of gaze, this requirement is fulfilled by adjusting the orientation of the prism to correctly position it perpendicular to the fixation target. It can no longer be held in the frontal plane of the patient and be perpendicular to the direction of the fixation object. In exotropia, lateral incomitance can be induced by allowing the prism to follow the rotation of the head.[18] For near measurement the posterior face of the prism is rotated inward slightly to better approximate the minimum deviation position for which it is calibrated.

Measurement Errors

Prism placement. The clinical importance of errors that can occur when a prism is malpositioned, as described by Hardy in 1945,[10;11] was disregarded for almost 40 years.

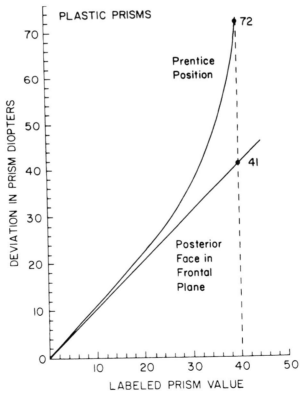

FIGURE 1-6. The deviation in prism diopters versus the labeled value of plastic prisms held in the Prentice and frontal plane positions.

From Thompson JT, Guyton DL: *Ophthalmology* 90:204-210, 1983.

It was not until the 1980s that ocular motility publications again began to underline its significance.[4;5;21;22] Although the errors are small for prism powers of less than 20Δ, when held in a position for which they are not calibrated, they increase significantly for larger prisms. For example, although a 40Δ power calibrated for the minimum deviation has a 41Δ power when held in primary position in the frontal plane, when held in the Prentice position, it has a power of 72Δ (Fig. 1-6). In a Bielschowsky head-tilt test an incorrectly positioned prism, not held parallel to the floor of the orbit or in line with the palpebral fissure, will induce an oblique prism effect and erroneous measurements. This effect can be readily observed by rotating a prism in front of one's eye in a straight-ahead gaze or by holding it parallel to the floor of the room while tilting the head and eye behind it.

Distance of prism from fixation target. Just as a plus lens, when moved away from the eye, increases its effective power and a minus lens decreases it, *prism effective power is altered by its position for near vision* (Fig. 1-7). For distance fixation there is no practical loss of effectivity. For near fixation a prism loses some effective power.[7;16;22] For example, a 15Δ prism located 25 mm in front of the center of rotation of the eye or 15 mm from the cornea (normal position) loses about 1Δ of effectiveness when the wearer reads at a distance of 33 cm. This negligible initial loss becomes more pronounced as the power of the prism is increased, as it is moved farther away from the eye, or as the object being viewed is moved closer. At 50 mm from the center of rotation, the loss for a 15Δ becomes 2Δ, and for a 50Δ it becomes 6.7Δ. The deviation will therefore be overestimated by this amount at near fixation, and this can be easily calculated by applying this simple formula.[7(pp19-120)]

$$R = \frac{P(D)}{D + d}$$

R = reduced effectivity
P = calibrated power of the prism
D = distance in meters of the fixation point from the prism
d = distance in meters of the prism from center of rotation of the eye

$$R = \frac{50(0.33)}{0.33 + 0.05}$$
$$R = 43.421$$

Risley rotary prisms in refractors or phoropters, calibrated for distance (see Risley prisms), are located several millimeters in front of the lens aperture or the normal spectacle plane. Therefore with higher powers they tend to overestimate phorias and fusional vergences at near.

When deviation measurements are done with prisms close to the eyes, in the position for which they are calibrated and for fixation targets at 6 m or more, such overestimation of large-angle deviations are minimized, and false vertical readings are avoided.

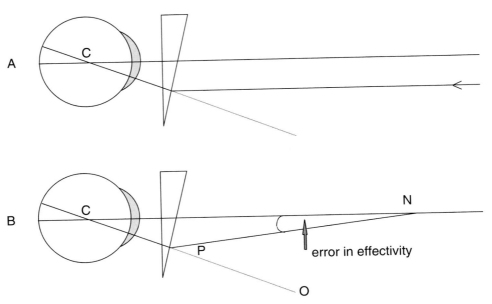

FIGURE 1-7. Error in effectivity. **A,** At distance, no loss of effectivity. Angle measured by prism = angle of rotation. **B,** At near, error increases as fixation object approaches eye. NPO (angle measured by prism) − CNP = NCO (angle of rotation).

Stacking of prisms. *Inaccurate measurement of a deviation will be obtained when prisms are "stacked" in the same direction* because the total deviation induced by two or more prisms held in series is greater than the sum of their labeled powers, resulting in an underestimate of the deviation.[5;21] For example, a 20Δ plastic prism stacked over a 30Δ plastic prism in the frontal plane position has a combined effective power of 66Δ (Table 1-1). This error results from the juxtaposition of the two prisms at their interface, which causes a ray of light passing through the first prism to be deviated to enter the second prism at a greater angle than its calibrated angle of incidence. After passing through the second prism, on emerging the light ray is deviated a still larger amount. Because it is impractical to hold separate prisms over the same eye according to their respective calibrated positions, it is better to avoid altogether stacking prisms in the same direction. Therefore for the examiner to determine precisely the residual angle of a patient's incorporated spectacle prismatic correction or a Fresnel membrane, when measurements require stacking additional prisms in the same direction, it is better to remove the spectacle prismatic correction. Otherwise the true deviation can be assessed only after a geometric-optical calculation of the correction.[5;21]

On the other hand, *a horizontal prism may be stacked with a vertical prism because the two vector components do not contribute to each other.*

Risley prisms. Risley prisms are composed of two identical right-angle prisms mounted back to back that can be contrarotated in relation to each other. Their maximum potential power corresponds to their added powers.

TABLE 1-1. Deviation in prism diopters for the addition of two plastic prisms stacked together, with the posterior prism in the frontal plane position.

Added prism (labeled value in prism diopters)	Initial prism (labeled value in prism diopters)											
	10	12	14	16	18	20	25	30	35	40	45	50
1	11	13	15	17	19	21	27	32	37	43	48	54
2	12	14	16	18	20	23	28	33	39	45	50	56
3	13	15	17	19	22	24	29	35	40	46	52	58
4	14	16	18	21	23	25	30	36	42	48	54	61
5	15	17	20	22	24	26	32	38	44	50	56	63
6	16	19	21	23	25	27	33	39	45	52	59	66
7	17	20	22	24	26	29	35	41	47	54	61	68
8	19	21	23	25	28	30	36	42	49	56	63	71
9	20	22	24	27	29	31	37	44	51	58	66	74
10	21	23	25	28	30	33	39	46	53	60	68	77
12	23	25	28	30	33	35	42	49	57	65	74	84
14	25	28	30	33	35	38	45	53	61	70	80	91
16	28	30	33	36	38	41	49	57	66	76	87	100
18	30	33	35	38	41	44	52	61	71	82	95	110
20	33	35	38	41	44	47	56	66	76	89	104	122
25	39	42	45	49	52	56	66	78	93	110	133	165
30	46	49	53	57	61	66	78	94	114	141	183	264
35	53	57	61	66	71	76	93	114	144	195	315	—
40	60	65	70	76	82	89	110	141	195	339	—	—
45	68	74	80	87	95	104	133	183	315	—	—	—
50	77	84	91	100	110	122	165	265	—	—	—	—

From Thompson JT, Guyton DL: *Ophthalmology* 90:204-210, 1983.

The induced vector effect is precisely calculated.[7p120] In changing the orientation of the two prisms, their resultant power varies to achieve the maximum power, once the vectors have become parallel. When their bases are directly opposed the added powers become zero (Fig. 1-8 and 1-9). The original Risley prism[19] was designed to fit readily in a trial frame. It opposed two 15Δ prisms and therefore had a range of powers from 0Δ to 30Δ. Today most refractors or phoropters are equipped with two rotary prisms with powers selected by the purchaser. They can be oriented to give either a horizontal prism position (Fig. 1-8) or a vertical prism position (Fig. 1-9) or both. The instrument is calibrated to give an accurate reading at 6 m. The reduced field and fixed position limit their use to the primary position. Risley prisms have the advantage of providing a smooth progressive increase of prism power and are preferred to prism bars by some examiners for the measurement of fusion amplitudes. The Risley prism is also available as a separate, hand-held or trial frame–supported unit. The hand-held unit facilitates its use for near measurements (see Fig. 3-1).

Splitting prism power between the two eyes. *Splitting the prism power between the two eyes significantly reduces prism-stacking errors,* although this still can induce underestimation of a deviation. However, for prisms up to 30Δ

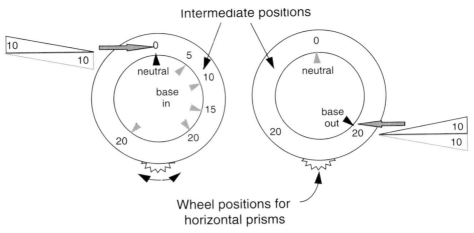

FIGURE 1-8. Phoropter Risley prisms: configuration for horizontal prisms. Usual position of knurled control wheels at 6 o'clock position for base-in and base-out. Alternative 12 o'clock position is less convenient for patient or examiner use. Black arrow indicates position of base of prism and resultant power, progressive from 0Δ to 20Δ for each eye.

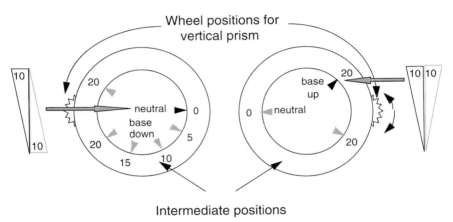

FIGURE 1-9. Phoropter Risley prisms: configuration for vertical prisms. Knurled control wheels have been rotated temporally for base-up and base-down. Alternative nasal position is awkward for manipulation. Black arrow indicates position of base of prism and its resultant power, progressive from 0Δ to 20Δ for each eye.

over each eye the error is small. For example, a 25Δ over each eye has a combined effective power of 53Δ (Table 1-2).

High-power spectacle lenses. Thick spectacle lenses not only induce prism measurement errors but also impede the observation of the eyes in lateral gaze directions. To be deemed accurate the prism and cover test findings should be in accord with those of versions and ductions.

A *plus lens* can be considered as two prisms juxtaposed base-to-base at its center. In an esodeviation when the eye is displaced inward in relation to the optical center of a plus lens, a base-out prism effect is induced; when displaced

TABLE 1-2. Deviation in prism diopters for the addition of two prisms (glass or plastic) with one prism held in front of each eye.

Left eye prism (labeled value)	Right eye prism (labeled value)											
	10	12	14	16	18	20	25	30	35	40	45	50
10	20	22	24	26	29	31	36	41	47	52	58	63
12	22	24	26	29	31	33	38	44	49	55	60	66
14	24	26	29	31	33	35	40	46	52	57	63	69
16	26	29	31	33	35	37	43	48	54	60	66	72
18	29	31	33	35	37	39	45	51	57	63	69	75
20	31	33	35	37	39	42	47	53	59	65	71	78
25	36	38	40	43	45	47	53	59	66	72	79	86
30	41	44	46	48	51	53	59	66	73	80	87	94
35	47	49	52	54	57	59	66	73	80	87	95	103
40	52	55	57	60	63	65	72	80	87	95	104	113
45	58	60	63	66	69	71	79	87	95	104	113	123
50	63	66	69	72	75	78	86	94	103	113	123	133

From Thompson JT, Guyton DL: *Ophthalmology* 90:204-210, 1983.

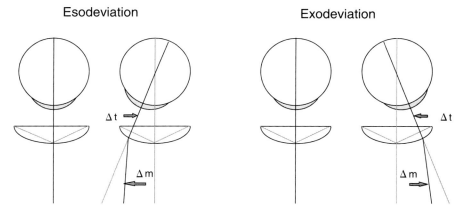

FIGURE 1-10. The effect of plus lenses on esodeviation and exodeviation. The measured deviation Δm is less than the true deviation Δt.

outward in an exodeviation, a base-in effect is generated (Fig. 1-10). Because less prism power is necessary to neutralize either deviation, they will be underestimated. A *minus lens* induces a base-in prism effect in an esodeviation and a base-out prism effect in an exodeviation (Fig. 1-11). Here more prism power is needed to neutralize either deviation, and they will be overestimated. With low-power lenses, the error is negligible. With high-power lenses, the true deviation can be found with the following formula[20]:

$$\left(\frac{\Delta T}{\Delta M}\right) \times 100 = \frac{100}{1 - 0.025\,D}\%$$

where ΔT represents the true deviation, ΔM is the measured deviation, and D is the dioptric power of the lens. The simplified formula $\Delta M \pm 2.5\% \times D$ is approximate. For example, a bilateral, -10.00 D myope with an exotropia mea-

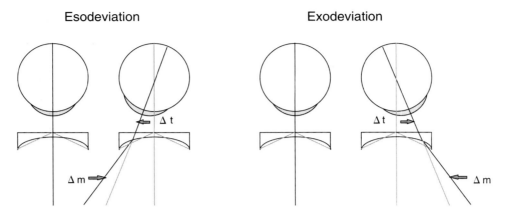

FIGURE 1-11. Effect of minus lenses on esodeviation and exodeviation. The measured deviation Δm is greater than the true deviation Δt.

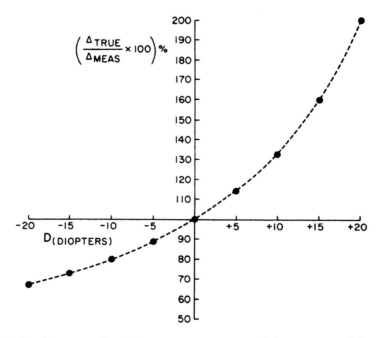

FIGURE 1-12. The true deviation as a percentage of the measured deviation for various powers of spectacle lenses from −20.00 to +20.00 diopters.

From Scattergood KD, Brown MH, Guyton DL: *Am J Ophthalmol* 96:439-448, 1983.

sured 40Δ at distance has a true deviation of 32Δ, and with the simplified formula (ΔM − 2.5% × D) a deviation of 30Δ. With higher powers the difference is more marked (Fig. 1-12).

A myopic correction apparently decreases the strabismus not only because of the minification induced but also because of the inverse prism effect. Conversely, a hyperopic correction, which magnifies and has a direct prism effect, makes a deviation more noticeable.

Compounding an Oblique Prism

A combined horizontal and vertical deviation can be corrected by a single prism with its apex placed in an oblique position. The appropriate power of the obliquely placed prism and its axis can be calculated by the vector formula or by simple trigonometry involving a right-angle triangle.

Power calculation. For example, if 15Δ is required to neutralize the horizontal deviation (H) and 10Δ is required to neutralize the vertical deviation (V), the power of the resultant (oblique) prism (P) corresponding to the hypotenuse of the triangle can be easily calculated with the Pythagorean theorem.

$$P^2 = V^2 + H^2$$
$$P = \sqrt{V^2 + H^2}$$
$$P = \sqrt{100 + 225}$$
$$P = 18\Delta$$

Angle calculation. The angle (Θ) at which this 18Δ oblique prism should be positioned is found from one of the following formulas:

$$\sin \Theta = \frac{V}{P}$$
$$\sin \Theta = \frac{10}{18}$$
$$\sin \Theta = 0.555 \Theta = 33.7°$$

or

$$\tan \Theta = \frac{V}{H}$$
$$\tan \Theta = \frac{10}{15}$$
$$\tan \Theta = 0.666$$
$$\Theta = 33.7°$$

When a precise correction is indicated, refer to tables developed from these formulas (Tables 1-3 and 1-4),[13] refer to a nomograph (Fig. 1-13), or use a hand calculator with *tan* or *sin* signs to determine the exact power and angle for positioning of an oblique prism. Such precision is necessary to manufacture a lens when a prescription specifies that both the horizontal and the vertical prisms be in the same lens. The refractive index of the material used and the optical correction must be taken into account in the incorporation of a compound prism. The calculation in a spectacle prescription is better left to the optical laboratory. The examiner need only indicate the horizontal and vertical prism powers for each eye, with the designated powers being divided equally between the two eyes to minimize the side effect of unequal magnification.

On the other hand, *oblique Fresnel prism membranes* can be readily and accurately applied in the office over the patient's spectacle lens without consulting a complicated formula or nomogram. Moreover, a Fresnel membrane

TABLE 1-3.   Resultant powers for prisms of 1 to 30 diopters*

Vertical prism required (in diopters) / Horizontal prism required (in diopters)

V\H	1	2	3	4	5	6	7	8	9	10	11	12	13	14	15	16	17	18	19	20	21	22	23	24	25	26	27	28	29	30
1	1	2	3	4	5	6	7	8	9	10	11	12	13	14	15	16	17	18	19	20	21	22	23	24	25	26	27	28	29	30
2	2	3	4	4	5	6	7	8	9	10	11	12	13	14	15	16	17	18	19	20	21	22	23	24	25	26	27	28	29	30
3	3	4	4	5	6	7	8	9	9	10	11	12	13	14	15	16	17	18	19	20	21	22	23	24	25	26	27	28	29	30
4	4	4	5	6	6	7	8	9	10	11	12	13	14	15	16	16	17	18	19	20	21	22	23	24	25	26	27	28	29	30
5	5	5	6	6	7	8	9	9	10	11	12	13	14	15	16	17	18	19	20	21	22	23	24	25	25	26	27	28	29	30
6	6	6	7	7	8	8	9	10	11	12	13	13	14	15	16	17	18	19	20	21	22	23	24	25	26	27	28	29	30	31
7	7	7	8	8	9	9	10	11	11	12	13	14	15	16	17	17	18	19	20	21	22	23	24	25	26	27	28	29	30	31
8	8	8	9	9	9	10	11	11	12	13	14	14	15	16	17	18	19	20	21	22	22	23	24	25	26	27	28	29	30	31
9	9	9	9	10	10	11	11	12	13	13	14	15	16	17	17	18	19	20	21	22	23	24	25	26	27	28	28	29	30	31
10	10	10	10	11	11	12	12	13	13	14	15	16	16	17	18	19	20	21	21	22	23	24	25	26	27	28	29	30	31	32
11	11	11	11	12	12	13	13	14	14	15	16	16	17	18	19	19	20	21	22	23	24	25	25	26	27	28	29	30	31	32
12	12	12	12	13	13	13	14	14	15	16	16	17	18	18	19	20	21	22	22	23	24	25	26	27	28	29	30	30	31	32
13	13	13	13	14	14	14	15	15	16	16	17	18	18	19	20	21	21	22	23	24	25	26	26	27	28	29	30	31	32	33
14	14	14	14	15	15	15	16	16	17	17	18	18	19	20	21	21	22	23	24	24	25	26	27	28	29	30	30	31	32	33
15	15	15	15	16	16	16	17	17	17	18	19	19	20	21	21	22	23	23	24	25	26	27	27	28	29	30	31	32	33	34
16	16	16	16	16	17	17	17	18	18	19	19	20	21	21	22	23	23	24	25	26	26	27	28	29	30	31	31	32	33	34
17	17	17	17	17	18	18	18	19	19	20	20	21	21	22	23	23	24	25	25	26	27	28	29	29	30	31	32	33	34	34
18	18	18	18	18	19	19	19	20	20	21	21	22	22	23	23	24	25	25	26	27	28	28	29	30	31	32	32	33	34	35
19	19	19	19	19	20	20	20	21	21	21	22	22	23	24	24	25	25	26	27	28	28	29	30	31	31	32	33	34	35	36
20	20	20	20	20	21	21	21	22	22	22	23	23	24	24	25	26	26	27	28	28	29	30	30	31	32	33	34	34	35	36
21	21	21	21	21	22	22	22	22	23	23	24	24	25	25	26	26	27	28	28	29	30	30	31	32	33	33	34	35	36	37
22	22	22	22	22	23	23	23	23	24	24	25	25	26	26	27	27	28	28	29	30	30	31	32	33	33	34	35	36	36	37
23	23	23	23	23	24	24	24	24	25	25	25	26	26	27	27	28	29	29	30	30	31	32	33	33	34	35	35	36	37	38
24	24	24	24	24	25	25	25	25	26	26	26	27	27	28	28	29	29	30	31	31	32	33	33	34	35	35	36	37	38	38
25	25	25	25	25	25	26	26	26	27	27	27	28	28	29	29	30	30	31	31	32	33	33	34	35	35	36	37	38	38	39
26	26	26	26	26	26	27	27	27	28	28	28	29	29	30	30	31	31	32	32	33	33	34	35	35	36	37	37	38	39	40
27	27	27	27	27	27	28	28	28	28	29	29	30	30	30	31	31	32	32	33	34	34	35	35	36	37	37	38	39	40	40
28	28	28	28	28	28	29	29	29	29	30	30	30	31	31	32	32	33	33	34	34	35	36	36	37	38	38	39	40	40	41
29	29	29	29	29	29	30	30	30	30	31	31	31	32	32	33	33	34	34	35	35	36	36	37	38	38	39	40	40	41	42
30	30	30	30	30	30	31	31	31	31	32	32	32	33	33	34	34	34	35	36	36	37	37	38	38	39	40	40	41	42	42

From Moore S, Stockbridge L: *Am Orthop J* 22:14-21, 1972.
*The intersection of the horizontal prism required and the vertical prism required gives the power for the oblique prism. For example, for an esotropia of 28Δ and a vertical deviation of 10Δ, a 30Δ prism would be required. (See Table 1-4 for the axis.)

TABLE 1-4. Angles for prisms of 1 to 30 diopters*

Vertical prism required (in diopters)	Horizontal prism required (in diopters)																													
	1	**2**	**3**	**4**	**5**	**6**	**7**	**8**	**9**	**10**	**11**	**12**	**13**	**14**	**15**	**16**	**17**	**18**	**19**	**20**	**21**	**22**	**23**	**24**	**25**	**26**	**27**	**28**	**29**	**30**
1	45	27	18	14	11	9	8	7	6	6	5	5	4	4	4	4	3	3	3	3	3	3	2	2	2	2	2	2	2	2
2	63	45	34	27	22	18	16	14	13	11	10	9	9	8	8	7	7	6	6	6	5	5	5	5	5	4	4	4	4	4
3	72	56	45	37	31	27	23	21	18	17	15	14	13	12	11	11	10	9	9	9	8	8	7	7	7	7	6	6	6	6
4	76	63	53	45	39	34	30	27	24	22	20	18	17	16	15	14	13	13	12	11	11	10	10	9	9	9	8	8	8	8
5	79	68	59	51	45	40	36	32	29	27	24	23	21	20	18	17	16	16	15	14	13	13	12	12	11	11	10	10	10	9
6	81	72	63	56	50	45	41	37	34	31	29	27	25	23	22	21	19	18	18	17	16	15	15	14	13	13	13	12	12	11
7	82	74	67	60	54	49	45	41	38	35	32	30	28	27	25	24	22	21	20	19	18	18	17	16	16	15	15	14	14	13
8	83	76	69	63	58	53	49	45	42	39	36	34	32	30	28	27	25	24	23	22	21	20	19	18	18	17	16	16	15	15
9	84	77	72	66	61	56	52	48	45	42	39	37	35	33	31	29	28	27	25	24	23	22	21	21	20	19	18	18	17	17
10	84	79	73	68	63	59	55	51	48	45	42	40	38	36	34	32	30	29	28	27	25	24	23	23	22	21	20	20	19	18
11	85	80	75	70	66	61	58	54	51	48	45	43	40	38	36	35	33	31	30	29	28	27	26	25	24	23	22	21	21	20
12	85	81	76	72	67	63	60	56	53	50	47	45	43	41	39	37	35	34	32	31	30	29	28	27	26	25	24	23	22	22
13	86	81	77	73	69	65	62	58	55	52	50	47	45	43	41	39	37	36	34	33	32	31	29	28	27	27	26	25	24	23
14	86	82	78	74	70	67	63	60	57	54	52	49	47	45	43	41	39	38	36	35	34	32	31	30	29	28	27	27	26	25
15	86	82	79	75	72	68	65	62	59	56	54	51	49	47	45	43	41	40	38	37	36	34	33	32	31	30	29	28	27	27
16	86	83	79	76	73	69	66	63	61	58	55	53	51	49	47	45	43	42	40	39	37	36	35	34	33	32	31	30	29	28
17	87	83	80	77	74	71	68	65	62	60	57	55	53	51	49	47	45	43	42	40	39	38	36	35	34	33	32	31	30	30
18	87	84	81	77	74	72	69	66	63	61	59	56	54	52	50	48	47	45	43	42	41	39	38	37	36	35	34	33	32	31
19	87	84	81	78	75	72	70	67	65	62	60	58	56	54	52	50	48	47	45	44	42	41	40	38	37	36	35	34	33	32
20	87	84	81	79	76	73	71	68	66	63	61	59	57	55	53	51	50	48	46	45	44	42	41	40	39	38	37	36	35	34
21	87	85	82	79	77	74	72	69	67	65	62	60	58	56	54	53	51	49	48	46	45	44	42	41	40	39	38	37	36	35
22	87	85	82	80	77	75	72	70	68	66	63	61	59	58	56	54	52	51	49	48	46	45	44	43	41	40	39	38	37	36
23	88	85	83	80	78	75	73	71	69	67	64	62	61	59	57	55	54	52	50	49	48	46	45	44	43	41	40	39	38	37
24	88	85	83	81	78	76	74	72	69	67	65	63	62	60	58	56	55	53	52	50	49	47	46	45	44	43	42	41	40	39
25	88	85	83	81	79	76	74	72	70	68	66	64	63	61	59	57	56	54	53	51	50	49	47	46	45	44	43	42	41	40
26	88	86	83	81	79	77	75	73	71	69	67	65	63	62	60	58	57	55	54	52	51	50	49	47	46	45	44	43	42	41
27	88	86	84	82	80	77	75	73	72	70	68	66	64	63	61	59	58	56	55	53	52	51	50	48	47	46	45	44	43	42
28	88	86	84	82	80	78	76	74	72	70	69	67	65	63	62	60	59	57	56	54	53	52	51	49	48	47	46	45	44	43
29	88	86	84	82	80	78	76	75	73	71	69	68	66	64	63	61	60	58	57	55	54	53	52	50	49	48	47	46	45	44
30	88	86	84	82	81	79	77	75	73	72	70	68	76	65	63	62	60	59	58	56	55	54	53	51	50	49	48	47	46	45

From Moore S, Stockbridge L: *Am Orthop J* 22:14-21, 1972.

*The intersection of horizontal and vertical prism powers required gives the angle at which the oblique prism (**see Table 1-3 for power**) is placed. Continuing the example, the intersection of 28Δ and 10Δ is 20°, the axis at which the 30Δ is placed to correct the two deviations.

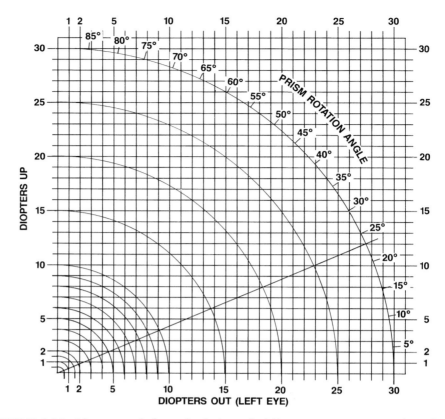

FIGURE 1-13. Nomograph for calculation of oblique prism power and angle.

From 3M Corporation, promotional handout St. Paul, MN, 1971.

Instructions for use:
1. Determine the amounts of vertical and horizontal prism required, and locate their intersection on the nomograph.
2. The quarter circle nearest this intersection is the proper total prism power.
3. Locate the intersection of this quarter circle and the exact vertical prism required and draw a diagonal line from the lower left corner through this intersection, continuing to the *prism rotation angle* scale. The angle shown on the scale is the correct orientation of the prism axis to obtain the exact vertical prism required.
4. The exact amount of horizontal prism obtained can be found by noting where the intersection of the quarter circle and the diagonal line falls on the horizontal scale.

> N.B. The procedure is designed to achieve exact vertical prism, while allowing fractional variations in horizontal prism.

of any available power induces only negligible magnification effect and therefore can be applied over one lens only.

Oblique Fresnel membranes are frequently indicated for the following patients:

1. Adults with vertical and horizontal diplopia who, as a rule, do not tolerate a Fresnel prism over both eyes.

2. Children for which correction of a vertical and horizontal deviation with a single membrane over one eye serves as a partial occluder or when over both eyes equalizes retinal stimuli.

The power and oblique-angle position of a prism can be readily determined clinically with the practical objective method that follows. Should the examiner refer to a table or nomogram, *the commercially available powers of Fresnel membranes have to be kept in mind*. It is of little use to determine that the patient requires an 18Δ oblique prism if no 18Δ Fresnel membrane is available. The chart of Reinecke et. al.[17] for deviations of 15Δ or less takes into account the membranes commercially available and can be used for quick reference (Table 1-5).

Compounding an Oblique Prism for a Fresnel Membrane

Practical objective method. A Fresnel prism with a power greater than the larger deviation but less than the combined errors found with the cover test is chosen from the trial set, and its apex is positioned halfway between the horizontal and vertical deviations. For example, in an adult patient or a child with an exotropia of 25Δ and a hypotropia of 10Δ, a 30Δ Fresnel prism is placed in a Halberg Clip-On over the patient's spectacles or inserted in a trial frame with its apex oriented temporally and inferiorly over the deviated hypotropic eye. The cover test is repeated as the prism is rotated until the deviation is neutralized. If there is a vertical overcorrection, the apex of the prism is rotated upward, toward the horizontal, to compensate. When the vertical deviation is fully corrected, if a residual horizontal deviation persists, the 30Δ Fresnel prism is replaced with a 35Δ. Once vertical and horizontal neutralization is achieved, prism power and axis are recorded.

For a child, where the prisms are to be divided about equally between the two eyes, the examiner starts with a 15Δ trial prism over each eye placed in a slightly oblique position as in the example given earlier, with *the apices of the prisms pointing in the direction of the deviation to be neutralized*. In the hypotropic eye, the apex is temporal and slightly down. In the hypertropic eye, the apex is temporal and slightly up. The cover test is repeated and the prisms are rotated or changed in turn until the deviation is neutralized.

With experience, *accurate prism power and orientation can be successfully determined in a few minutes*. The advantage of this technique is that a Fresnel trial prism finding can be transposed directly to a membrane of corresponding power for the immediate final application.

Subjective method for patient. When an oblique prism is prescribed to correct diplopia, the same approach is used, but instead of doing a cover test, the trial prism is rotated to subjectively compensate the diplopia. In this instance it is often not necessary to fully correct the vertical and horizontal deviations to overcome diplopia. A membrane with the required power and orientation is applied to the patient's spectacle lens over the nonpreferred eye.

Oblique prism notation. By convention the degree notations on a trial frame are recorded counterclockwise from 0° to 180° for the higher and lower semi-

TABLE 1-5. Chart of resultant prism power and angle of orientation for given combinations of horizontal and vertical prism power requirements.

Horizontal prism required (in diopters)

Vertical prism required (in diopters)	1	2	3	4	5	6	7	8	9	10	11	12	13	14	15
1	2△/30°	2△/30°	3△/19°	4△/14°	5△/12°	6△/10°	7△/8°	8△/7°	9△/6°	10△/6°	12△/5°	12△/5°	15△/4°	15△/4°	15△/4°
2	2△/84°	3△/42°	4△/30°	4△/30°	5△/24°	6△/19°	7△/17°	8△/14°	9△/13°	10△/12°	12△/10°	12△/10°	15△/8°	15△/8°	15△/8°
3	3△/84°	4△/49°	4△/49°	5△/37°	6△/30°	7△/25°	8△/22°	9△/19°	9△/19°	10△/17°	12△/14°	12△/14°	15△/12°	15△/12°	15△/12°
4	4△/84°	5△/53°	5△/53°	6△/42°	6△/42°	7△/35°	8△/30°	9△/26°	10△/26°	10△/24°	12△/19°	12△/19°	15△/15°	15△/15°	15△/15°
5	5△/84°	6△/56°	6△/56°	7△/46°	7△/46°	8△/39°	9△/34°	9△/34°	10△/30°	12△/25°	12△/25°	12△/25°	15△/19°	15△/19°	15△/19°
6	6△/84°	7△/59°	7△/59°	7△/59°	8△/49°	9△/42°	9△/42°	10△/37°	10△/37°	12△/30°	12△/30°	12△/30°	15△/24°	15△/24°	15△/24°
7	7△/84°	7△/84°	8△/61°	8△/61°	9△/51°	9△/51°	10△/44°	10△/44°	12△/36°	12△/36°	12△/36°	15△/28°	15△/28°	15△/28°	15△/28°
8	8△/84°	8△/84°	9△/63°	9△/63°	9△/63°	10△/53°	10△/53°	12△/42°	12△/42°	12△/42°	15△/32°	15△/32°	15△/32°	15△/32°	15△/32°
9	9△/84°	9△/84°	10△/64°	10△/64°	10△/64°	10△/64°	12△/49°	12△/49°	12△/49°	15△/37°	15△/37°	15△/37°	15△/37°	15△/37°	20△/27°
10	10△/84°	10△/84°	10△/84°	12△/56°	12△/56°	12△/56°	12△/56°	12△/56°	15△/42°	15△/42°	15△/42°	15△/42°	15△/42°	15△/42°	20△/30°
11	12△/66°	12△/66°	12△/66°	12△/66°	12△/66°	12△/66°	12△/66°	15△/47°	15△/47°	15△/47°	15△/47°	15△/47°	15△/47°	20△/33°	20△/33°
12	12△/85°	12△/84°	12△/84°	12△/84°	15△/53°	15△/53°	15△/53°	15△/53°	15△/53°	15△/53°	15△/53°	15△/33°	15△/47°	20△/37°	20△/37°
13	15△/60°	15△/60°	15△/60°	15△/60°	15△/60°	15△/60°	15△/60°	15△/60°	15△/60°	15△/60°	15△/60°	20△/37°	20△/37°	20△/41°	20△/41°
14	15△/69°	15△/69°	15△/69°	15△/69°	15△/69°	15△/69°	15△/69°	15△/69°	15△/69°	20△/44°	20△/44°	20△/44°	20△/44°	20△/44°	20△/44°
15	15△/86°	15△/84°	15△/84°	15△/84°	15△/84°	15△/84°	20△/49°	20△/49°	48△/49°	20△/49°	20△/49°	20△/49°	20△/49°	20△/49°	20△/49°

From Reinecke RD et al: *Arch Ophthalmol* 95:1255-1257, 1977.

*For example, a requirement of a 10-diopter horizontal prism and a 3-diopter vertical prism indicates 10△/17° (i.e., a 10-diopter prism with its base set at an angle of 17°). Cells with 4△ or greater horizontal error are indicated by underlines.

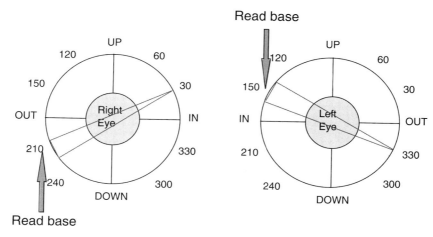

FIGURE 1-14. Prism notation in degrees and base position on a 360° circle divided into four quadrants.

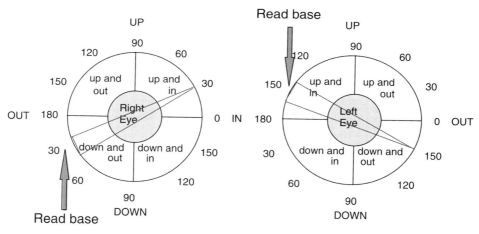

FIGURE 1-15. English system: prism notation defined in degrees using 180° hemicircles and by base position up or down, in or out.

circles for both the left eye and the right eye. Because the cylinder axis of a spectacle lens extends across the entire lens, its position can be properly specified by referencing only the upper halves of the circles. However, the base of a prism can assume any position on a circle from 0° to 360°. When the full circle notation is adopted (Fig. 1-14), the 0°, 90°, 180°, and 270° positions are deleted in favor of the in, out, up, and down base-position notation. The English system, by indicating also that the base of the prism is in and up, up and out, down and out, or down and in, can keep the 0° to 180° notation (Fig. 1-15).

The method I prefer is to follow the familiar 180° cylinder notation, *indicating that either the apex or the base of the prism is positioned in one of the two upper quadrants.* For example, a prism with its base at 210° is identified by its apex at 30° (Fig. 1-16).

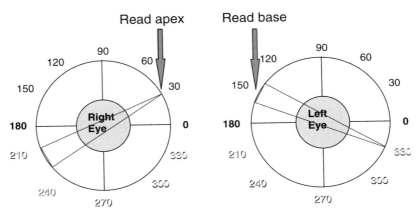

FIGURE 1-16. Simplified system. Apex or base read on upper half of trial frame markings in degrees.

Resolving Prisms

Manual lensmeters (lensometers). *The major reference point* of a lens with a prismatic correction is the point manifesting the amount and direction of the prism called for in the prescription that also incorporates the proper dioptric power.[9] It should coincide with the visual axis, which is marked on the lens, while the patient gazes at a fixation point at 6 m for the distance vision spectacles and at 33 cm for the reading spectacles. Afterward, with a lensmeter or similar instrument the examiner checks that the point marked on the lens coincides with the major reference point. Prisms incorporated in spectacles, whether induced by decentration or ground in, can have their power, up to 6Δ, and direction determined accurately with most lensmeters. The lensmeter targets in manual lensmeters are seen displaced in the direction of the base of the prism (Keplerian telescope). The targets are imaged on calibrated crossbars (Zeiss type) (Figs. 1-17 and 1-18) or on a calibrated set of rings (Bausch and Lomb type) (see Figs. 1-19 and 1-20) in 0.5Δ or 1Δ steps from which the dioptric power is read (up to 6Δ). When the prism is incorporated in a plano lens, centration of a prism-bearing lens is not necessary because the displacement remains constant from wherever the prism power is read within the lens diameter. However, *in a powered lens, because of the prismatic effect induced by decentration (see Fig. 1-22) the lens must be exactly centered as well as focused for an accurate reading of an incorporated prism power.* In the Zeiss lensmeter with a circular dot-pattern target and calibrated crossbars (Figs. 1-17 and 1-18), the power of the prism, its orientation in degrees, and the base direction can be determined simultaneously by rotating the calibrated cross (meridian indicator) so that the reticule scale intersects the center of the focused target pattern. If the base-apex line of the prism coincides with the axis of a simple cylindric lens, as with a plano lens, its power will not be altered by decentration (see later section, Prism Effect of Lens Decentration).

In a Bausch and Lomb lensmeter with the grid target pattern (Figs. 1-19 and 1-20), when the target has been focused at the lens power of an exactly

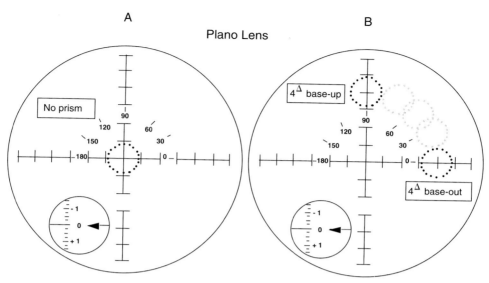

FIGURE 1-17. A, Lensmeter target having a circular dot pattern. **B,** Resolving a prism. The power of the prism and its axis are read directly. The target displacement is toward the base. Its orientation is read along the 0° to 180° hemicircle.

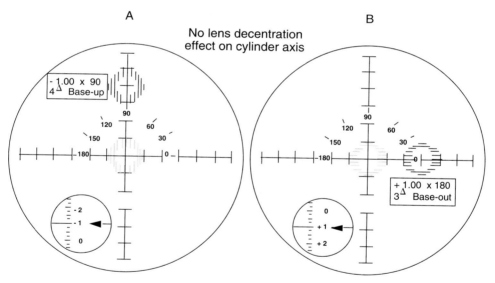

FIGURE 1-18. A, Resolving a vertical prism on the axis of a cylinder axis 90°. **B,** Resolving a horizontal prism on the axis of a cylinder axis 180°. The elongated target has been focused, indicating the cylinder power and axis. In both examples the prism had to be incorporated (ground in).

centered lens with the right-hand knob, the eyepiece, fitted with a meridian indicator line, is rotated to coincide with the intersection of the perpendicular center lines of the 3/1 target pattern on the graduated ring. The intersected ring indicates the prism power in the direction of the base-apex line. Then the focused target pattern is rotated with the left knob (calibrated in degrees)

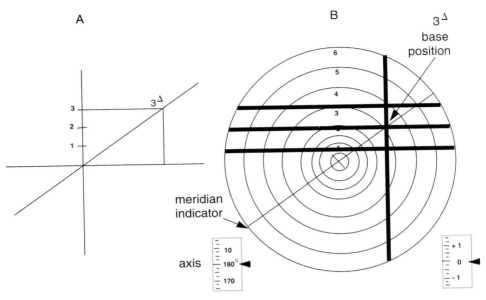

FIGURE 1-19. Resolving an oblique prism. **A,** By a parallelogram. **B,** Using a grid target lensmeter.

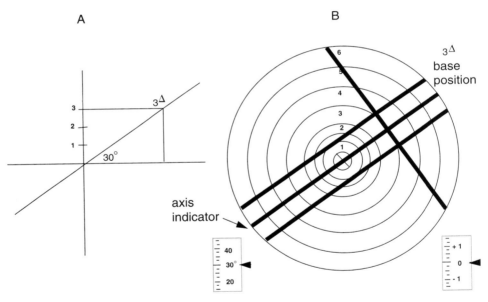

FIGURE 1-20. Determining the axis of an oblique prism. **A,** By a parallelogram. **B,** Using a grid target lensmeter.

so that the central of the three target lines is superimposed on the meridian indicator, and the orientation (axis) is read in degrees (Fig. 1-20).

Automated lensmeters (lensometers). The automated lensmeters are fitted with a microcomputer that calibrates and processes the data from a light target. Centration is easily accomplished. The spherical, cylindric, and prism powers (up to 10Δ) are automatically read, and the data values are printed out. The displacement of the prism target can be in the direction of the base or the apex, depending on the instrument design.

By eliminating the manual adjustment of the eyepiece and subjective calculations and by providing easy centration of the lens, readings are more quickly and accurately obtained than with manual instruments. Nevertheless, to eliminate human as well as machine errors it is wise to check the accuracy of a given instrument with a spherocylinder and prism lens combination of known power before relying on it.

The reading of prism powers above the limits that can be determined by either manual or automated lensmeters can be supplemented with loose trial prisms after compensating for the spherocylindric power of the spectacle.

Decentration is customarily used only to achieve a small prism correction. It can be readily determined that a prismatic effect has been achieved by grinding rather than by decentering when the greatest distance from the optical center of the lens is more than one-half the diameter of the usual 60-mm lens blank.[1(p56)]

A *Fresnel prism membrane* also must be neutralized at the optical center of a powered spectacle carrier lens to avoid the error of adding or subtracting prism power induced by lens decentration.

In powers above the range of the lensmeter, a Fresnel membrane, especially in higher powers or when applied over a strong spectacle correction, is often more readily and accurately neutralized if detached from the lens. Remembering that an object is displaced toward the apex of a prism, the examiner first identifies the apex by looking with one eye through the membrane at the vertical plane of a distant object. The prism power is determined by using successive neutralizing prisms, their bases being placed in the opposite direction, up to the point the displaced object is centered. Care must be taken to hold both prisms in front of the same eye with one face perpendicular to the line of sight (Prentice position). Because the estimation is approximate, it is beneficial to remember the available powers of prism membranes.

Optical bench. A precise reading can be obtained with an easily assembled optical bench (Fig. 1-21). A helium-neon (HeNe) laser (pointer) has its collimated light striking a prism that is placed into the position for which it is calibrated, for example, striking perpendicular to one face of the prism (Prentice position). The deviation induced is read on a meter ruler placed at 1 m from the refracting surface of the prism. The ruler can be rotated on a protractor to determine prism power and axis in oblique positions. According to Prentice's rule, each centimeter of displacement corresponds to 1Δ. The variation of power with changes from the calibrated position can be readily observed. Conversely, for a prism of known power, its calibrated position can be determined.

Resolving oblique prisms. The power of an oblique prism can be resolved into its vertical and horizontal components in a fashion similar to its original compounding. The axis of the prism, represented by the diagonal of a paral-

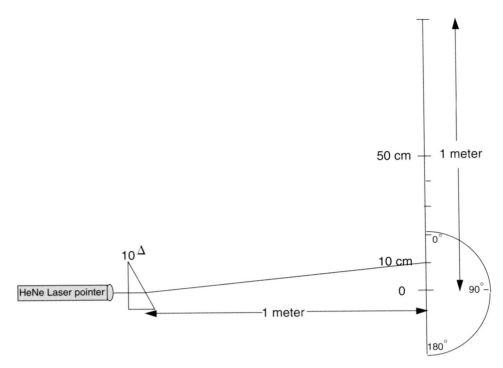

FIGURE 1-21. Schematic of an optical bench for measuring prism power. Columnated light strikes the prism in its calibrated Prentice position.

lelogram, is drawn on a coordinate system. Horizontal and vertical lines are dropped from it and measured (see Figs. 1-19 and 1-20). When a lensmeter with a crosspattern is used, the intersection of the sphere and cylinder lines gives the location of the base of the prism (see Fig. 1-19). Perpendiculars from this point to the lines of the inner scale (corresponding to x and y axes) give horizontal (H) and vertical (V) prism components. The eyepiece of the instrument must be rotated so that the reticule scale intersects the center of the perpendicular line (or dot pattern when a dot target is used) (see Fig. 1-17). Each unit of displacement on the scale represents 1Δ.

The resolving of oblique prism powers can also be calculated trigonometrically with the equations mentioned to compound it.

With P being the power of the oblique prism and Θ, the axis orientation,

$$V = P \sin \Theta$$

$$H = P \cos \Theta$$

Prism Effect of Lens Decentration

Decentration is used both to create a prismatic effect when a prismatic correction is desired and to compensate the unwanted prismatic effect induced by lenses of different powers in both eyes.

Decentration of a lens is defined as the displacement of the pole of a lens (optical center) from its geometric center.

A spectacle lens is designed so that its optical axis passes through the center of the pupil, or more precisely to coincide with the line of sight. The prismatic effect (Δ) induced by lens decentration depends on the amount of decentration in centimeters (c) and the power of the lens (D). *By applying Prentice's rule this can be easily calculated.*

$$\Delta = c \times D$$

Decentration is used when the prismatic effect required is not large and when the power of the lens is sufficiently great so that only a small decentration is needed. When decentration cannot provide the desired prismatic effect, prism power may be ground in the lens by surfacing procedures. While the prism is ground into a lens, the surface power and thickness must be preserved to provide the correct back vertex power.

Spherical lenses. Because a *plus lens* may be considered a concentric series of prisms placed base to base from the center of a lens, *the direction of the decentration will coincide with the base of the prism.*

A *minus lens* may be considered a concentric series of prisms with apices directed toward the center and bases directed toward the periphery. *The direction of the decentration is opposed to the base of the prism* (Fig. 1-22).

For example, a patient is wearing a +6.00 spherical lens over both eyes, and the prescription calls for a 2Δ base-out for each eye. The amount of decentration (c) required is found by applying Prentice's rule.

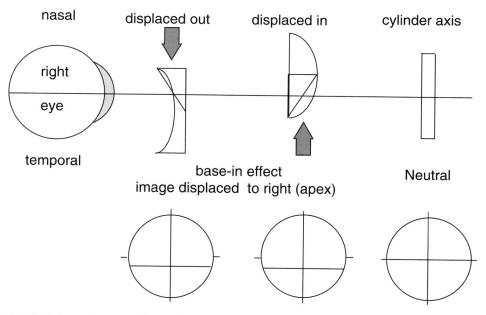

FIGURE 1-22. Prism effect of lens decentration over right eye: base-in effect with minus lens displaced temporally and with plus lens displaced nasally. No displacement induced along cylinder axis.

$$c \times 6 = 2\Delta$$

$$c = \frac{2}{6} = 0.33 \text{ cm}$$

To achieve a 2Δ base-out effect the $+6.00$ lens must have its optical center displaced temporally 3 mm from the center of the pupil. A -6.00 spherical lens used to achieve the same effect will have to be displaced 3 mm nasally.

Cylindric lenses. *The prismatic effect of a decentered cylindric lens is perpendicular to the cylinder axis.* A cylindric lens may be decentered to create a prismatic effect only when the prescribed base direction coincides with the direction of the power meridian. For example, a cylinder with a horizontal axis can be decentered only to produce a vertical prismatic effect (base-up or base-down), whereas a cylinder with a vertical axis can be decentered only to produce a horizontal prismatic effect (base-in or base-out), recalling the direction of decentration with plus and minus lenses to induce the desired effect (see Fig. 1-18). Prentice's rule also applies in the calculation of the displaced power meridian of a cylindric lens in any direction that is not on the cylinder axis.[7(p110)]

Change in gaze direction. A person who uses spectacles adjusted for the interpupillary distance in far gaze will experience at near a base-out prismatic effect with plus lenses and a base-in effect with minus lenses (Fig. 1-23). A prismatic effect will also be induced when looking above or below the optical center of the lenses. When a person looks down, a plus lens will induce a base-up effect (Fig. 1-24), and a minus lens will induce a base-down effect and inversely when gazing upward. The *vertical* prismatic effect is of importance only if different for the two eyes. If the relative difference is more than 1Δ, it is usually corrected by decentering the lenses or by a "slab-off."

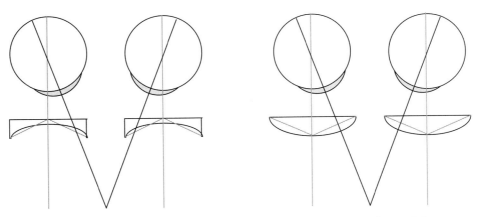

FIGURE 1-23. Induced prismatic effect when looking at near through lenses centered for distance: *(left)* base-in for minus lenses, and *(right)* base-out for plus lenses.

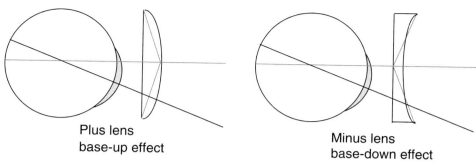

Plus lens
base-up effect

Minus lens
base-down effect

FIGURE 1-24. Induced prismatic effect when looking below the optical center of the lens.

Chromatic Aberration and Distortion of Prisms

Chromatic aberration. Short light waves (blue range) are more bent through a prism toward its base than are the longer light waves (red range). This phenomenon, termed *dispersion* or *chromatic aberration,* allowed Newton in his famous experiment to split the spectral components of white light into the various colors according to their wavelengths. For low powers used for incorporated prisms, chromatic aberration is not a major problem. However, dispersion increases with the power of the prism and is responsible for the color fringes seen with higher power prisms. Two prisms of different powers made of different materials with their bases oriented in opposite directions and two faces in direct contact could produce an achromatic prism. However, their combined weight and thickness would make such a prism impractical.[7(p152)] Over a modest band of wavelengths the chromatic aberration of a binary lens will cancel that of a conventional lens into which it has been etched (see Fig. 2-19).[25;26]

Distortion. In addition to the overall angular magnification of an ophthalmic prism, a change in the shape of the image will occur when the incident light is not perpendicular to the principal plane of the prism.[28] The apical angle created by the oblique incidence is greater than the calibrated apical angle of the prism, causing a greater deviation and distortion. The distortion depends not only on the power of the prism but also on its distance from the eye and its orientation in front of the eye. There is an increase in magnification toward the apex of the prism (Fig. 1-25). Images appear progressively stretched out in the base-apex meridian and even more toward the apex. These distortions are exaggerated with a change in the viewing angle. For example, with a horizontal prism the horizontal parallel lines appear to diverge as one looks toward the apex of the prism, whereas they converge when one looks toward the base. Also, lines perpendicular to the base-apex meridian appear concave toward the apex and convex toward the base.[14] With low-power prisms these distortions may pass unnoticed. However, with high powers they can cause spatial disorientation. Moreover, both vertical and horizontal

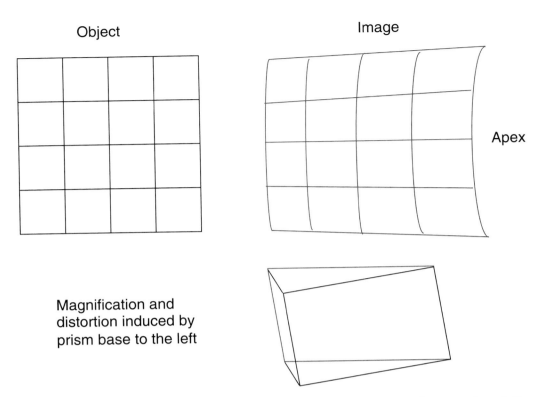

Object

Image

Apex

Magnification and
distortion induced by
prism base to the left

FIGURE 1-25. Image appears progressively stretched out with concavity of distortion toward the apex.

magnifications increase proportionally with an increasing base curve of the optical correction to which the prism is incorporated.[14]

If the power of the prismatic correction to be incorporated is divided equally between the two eyes, magnification will be better tolerated. Chromatic aberration and distortion induced by Fresnel prism membranes are discussed in Chapter 2.

Incorporated Prism Prescription

Bilateral prisms. When an incorporated prism is prescribed, it is customary to divide the power of horizontal or vertical prisms, or both, equally between the two eyes to prevent intolerable aniseikonia and to equalize the weight and thickness of the two finished lenses. If the optical correction between the two lenses differs, a slightly stronger prism can be prescribed on the side of the weaker dioptric correction after an office trial. The amount and direction of the prism called for in the prescription, or the major reference point, should fall directly in front of the center of the pupil and have the correct dioptric power as well.[9] Even when the power of the prismatic correction is divided equally between the two eyes, the patient may take time to adapt to the overall magnification the prisms induce, especially if this incorporation coincides with an increase of the hypermetropic correction.

Although incorporated prisms of higher powers are rarely prescribed, except in spectacle magnifiers, a highly motivated patient will tolerate good quality, properly adjusted prisms up to 15Δ over each eye for a horizontal deviation and up to 10Δ over each eye for a vertical deviation. The patient learns with time to move the head with the eyes to minimize distortions.

For powers of less than 8Δ incorporated in each lens, bifocals do not represent an added difficulty. Technically, different prismatic corrections can be incorporated in both lower and upper segments of a pair of single-vision lenses or bifocals, but they are unsightly or expensive. Should the patient require a different prismatic correction for distance and near and prefer incorporated prisms to Fresnel membranes, separate pairs of glasses for near and for distance are prescribed.

It is customary to indicate the position of the prism in a prescription by the direction of its base. As the apex always points in the direction of the deviation to be corrected (except for an inverse prism) the base will be oriented in the opposite direction. For example, for a right hypertropia necessitating an 8Δ correction, a 4Δ base-down prism will be prescribed over the right eye and a 4Δ base-up prism prescribed over the left eye. If the patient also needs a horizontal prism, the prescription will simply indicate its position base-in or base-out with its power divided equally without regard for the resulting oblique prism (see earlier section, Compounding an Oblique Prism for a Fresnel Membrane). Although conventional mathematics holds that for horizontal prisms a base-out position is positive and a base-in is negative, or that for vertical prisms a base-down position is positive and a base-up position is negative, this terminology is rarely used clinically.

Spectacle magnifiers. Spectacles with plus lenses as strong as 18 diopters and incorporating 18Δ base-in prism in each eye have been successfully prescribed to patients with subnormal vision. The magnification improves the near vision while the prisms prolong the reading period without asthenopia. As the reading material has to be placed at the anterior focal point of the correction the reading distance is significantly shortened, necessitating a greater effort of convergence with near accommodation. At a reading distance of 10 cm, with 10-meter angles, approximately 60Δ of convergence is required at an interpupillary distance of 60 mm (Fig. 1-26).[15] The "neutralizing" base-in prism can be obtained by decentration of the bifocal segments or by incorporated prisms in single-vision lenses. For bifocals, Fonda's rule-of-thumb is to decenter each segment 1 mm for each diopter of reading addition and in single-vision lenses to grind 1Δ of prism base-in for each diopter of reading addition in each eye. Decentration is impractical in single-vision, high plus lenses because the diameter of the lens blank is not large enough to provide for the large prismatic decentration requirement.[8]

Single direct and inverse prisms. A low-power incorporated prism may have to be prescribed for only one lens when a mechanical factor is present that induces symptoms or a compensatory head posture that is corrected only

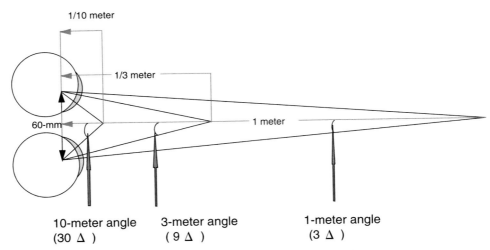

FIGURE 1-26. Total prism convergence requirement at 1-meter angle = 6Δ; at 3-meter angle (33 cm) = 18Δ; and at 10-meter angle (10 cm) = 60Δ.

when the total prismatic power is applied over the affected eye. If the patient does not accept wearing a Fresnel membrane, an incorporated prism of corresponding power is prescribed after office trial with a standard loose prism. With a unilateral prism under 6Δ, the difference in image size (aniseikonia) induced is usually well tolerated.

An inverse prism, that is, with its base or apex placed opposite to the anticipated direction, can be prescribed to improve the appearance of a poorly aligned prosthesis or of a deviated eye with poor vision for which surgical alignment is refused or not indicated. This prism can be incorporated in a magnifying plus lens or a minifying minus lens. The thickness and weight limit such cosmetic prescriptions to low-power corrections.

Another indication for inverse prisms would be the rarity of a patient with intractable diplopia, which is made more tolerable by further displacing the diplopia images. Inverse prisms in the treatment of heterophoria, amblyopia, and abnormal retinal correspondence (ARC) are discussed in Chapter 5.

When incorporated prisms are prescribed for either adults or children, the spectacle frame selected must be sturdy enough to hold thick lenses. A small round frame is advisable to reduce lens thickness and weight. The edge of the lenses are tinted the color of the frame to make the thickness less noticeable (see Chapter 2).

Accentuation of cosmetic blemish. Therapeutic prisms, especially high-power and vertical prisms, accentuate the cosmetic blemish of a deviation because the observer views the eye behind a prism to be displaced toward the apex of the prism in the same direction as the deviation. Parents and patients should be informed of this effect. Otherwise they will conclude that the condition is worsened by wearing prisms.

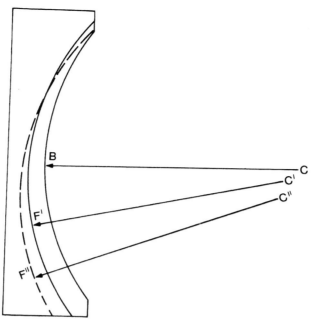

FIGURE 1-27. A 5Δ prism contact lens of +0.25 power; CB = 43.00 base curve cut on center; C'F' = front curve with center offset at C'. The displacement induces a prism of 5Δ. The offset C"F" has reached its practical limit, the upper quadrants have a knifelike edge.

From Véronneau-Troutman S: *Trans Am Ophthalmol Soc* 77:181-190, 1979.

Prisms in Contact Lenses

Prism ballast. To prevent a toric contact lens from rotating on the eye, a base-down prism can be incorporated in the lens. The increased thickness and weight of the lower part keeps the lens from too much rotation. This process, called prism ballast, can be used in hard or soft lenses.[7(p435)] It is also used for high-riding rigid gas permeable contact lenses that lend themselves to special applications and are not comfortable and functional otherwise.

The prism ballast does not usually exceed 2Δ and *is incorporated in both lenses* to neutralize induced secondary vertical base-down prismatic effect.

Prisms to overcome diplopia. There is little mention in the literature of incorporating a prism in a contact lens to compensate diplopia.[27] A contact lens with prism weight added in two adjacent quadrants will rotate to a position base-down, displacing downward the optical center of the eye. Consequently, it is limited to the compensation of vertical strabismus, and for this purpose, the prism is ground into one contact lens only.

The general formula to determine the difference in thickness of the base and the apex for a prism of given power and given size of a lens is:

$$T = \frac{DP}{100(n - 1)}$$

T = difference in thickness between base and apex (mm)
D = diameter of the lens (mm)
P = power of prism (Δ)
n = index of refraction of material used

In practice, the front curve of the lens is altered by a predetermined amount to produce the prismatic effect. This offset of the front curve of the lens has its practical limitations. A point is reached where the lens thickness in the upper two adjacent quadrants becomes a knifelike edge, too sharp and too fragile to be used (Fig. 1-27). The amount of prism and optical correction required are the principle factors contributing to the weight of the finished lens, its centration, and its tolerance. In theory, a high myopic contact lens can have a base-down effect induced by decentration without increasing much its center thickness. The thicker the center of the lens, as for the correction of high degrees of hyperopia, the heavier the lens becomes. Though a reduction in the overall diameter of such a lens reduces its weight, its fitting may be impaired. Two additional processes are used to further reduce the weight of the lens, thus improving its centration and tolerance.

Lenticularization. The lens is thinned in the same fashion as in a lenticular spectacle lens having a central optical zone and a surrounding nonoptical flange (Fig. 1-28). The purpose of the lenticular lens is to reduce the thickness without altering the diameter of the lens.

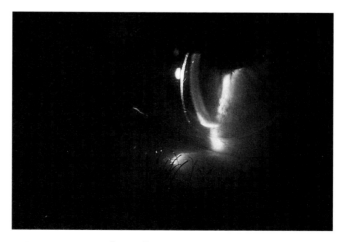

FIGURE 1-28. Prism contact lens showing lenticularization to reduce lens thickness to improve comfort and fit without altering power or diameter.

From Véronneau-Troutman S: *Trans Am Ophthalmol Soc* 77:181-190, 1979.

Feathering. The anterior peripheral aspect of the finished lens can be ground off, taking care to retain sufficient edge thickness to ensure a round, comfortable edge.

Truncation of the lens or total removal of the inferior periphery has no practical value. Although truncation has been tried to reduce the weight of the lens, the angulated base is thicker, causing a foreign-body sensation on the lower lid.

The accuracy of the prismatic effect is confirmed with the lensmeter and clinically. My experience has been limited to contact lenses correcting refractive errors of less than 3 diopters and incorporating ground-in, base-down prisms of no more than 6Δ. In theory, with binary optics, higher prism powers in other axis orientation could be incorporated into a very thin contact lens stabilized by a selectively positioned peripheral sector ballast.

Laser-Induced Corneal Prism (ΔPRK)

Corneal refractive surgery using the *argon fluoride* (ArF_2) *excimer laser* has made rapid progress during the past few years. This technology, called photorefractive keratectomy (PRK), is currently being used investigatively to remove graded amounts of tissue from the anterior cornea in a symmetric or meridional pattern to correct myopia, astigmatism, and, to a limited extent, hyperopia.

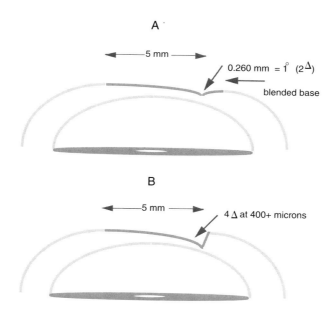

FIGURE 1-29. Schematic representations of anterior ArF_2 excimer laser corneal ablation forming low-power conventional prisms. **A,** Anschütz's Conoid laser keratectomy (CLK) with a 0.26 mm-cut, achieving 2Δ effect. Blending of the base of the prism to the corneal surface to promote more regular epithelialization. **B,** Azar's Prismatic photokeratectomy (PPK) with deeper cut, 0.4 mm, achieving a 4Δ base-up.

Conventional prism. In 1990 Anschütz[2] reported the use of the Aesculap MQL 60 laser and special templates to form a prismatic profile on polymethylmethacrylate test buttons and on pig eye corneas, which he termed *Conoid laser keratectomy* (CLK). To prevent pooling of tears and a potential site for development of necrosis and ulceration, he blended the abrupt, dellenlike, transition zone at the base (Fig. 1-29, A). He suggested that this modality might be useful to permanently compensate small extraocular muscle imbalances within the theoretical limits of correction obtainable. However, a 5-mm beam width required an ablation depth of 260 microns to obtain 1° (2Δ). By narrowing the stripe to 2 mm he could obtain this power at 60 microns. Nevertheless, because of the depths of ablation required (300 microns) to obtain higher prism powers, 5° (10Δ) and the corneal hyperopia induced by the central flattening (3 to 4 diopters at 100 microns calculated) he did not follow-up on his concept.

Azar[3] in 1993 presented his results using the 193-nm ArF_2 laser prismatic photokeratectomy (PPK) on polymethylmethacrylate (PMMA) blocks and PMMA model eyes and on rabbit eyes. He used a specially designed mask to form the base-up prismatic ablations at a diameter of 5 mm at depths of 0.060 mm, 0.120 mm, or 0.240 mm. Histologic examination at 9, 121, or 150 days showed reepithelialization and keratocyte responses only slightly delayed from that in control eyes subjected to standard photorefractive keratectomy myopia ablations. At deeper ablation levels, epithelial hyperplasia and corneal scarring were increased. He predicted a maximum power of 5Δ could be obtained

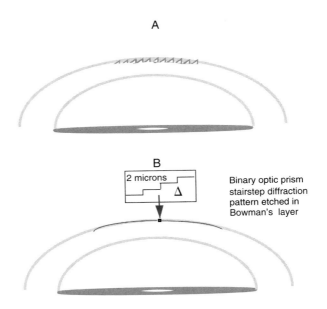

FIGURE 1-30. Schematic representations of anterior laser corneal ablation forming prisms of uniform thickness (ΔPRK). **A,** Applying the Fresnel principle. **B,** Applying binary optics with "microscopic" stairstep diffraction pattern.

in human eyes that, if formed on each eye, could be used to correct a total of 10Δ in consecutive small-angle strabismus, and postsurgical and postparetic binocular diplopia in adult patients (Fig. 1-29, B). Unlike Anschütz had calculated, Azar did not observe a change in corneal power toward hyperopia in the rabbit model.

Transposed to the human cornea, to achieve a 5Δ power at a 5-mm diameter, an ablation would have to be close to its full thickness at the base or, as Anschütz suggested, a smaller width would be required. The smaller width would significantly increase the potential for decentration, whereas a greater depth could result in delayed healing with persistent corneal haze. In time the thin base area eccentric to the optic axis could become ectatic, resulting in a decentered myopic (steeper) central cornea with irregular astigmatism.

ΔPRK Fresnel prism and ΔPRK binary optic prism. Because of the potential drawbacks of forming a full single-prism profile in the cornea, Troutman[24] and I propose to use the principle of the Fresnel prism membrane or of its extension binary optics (see Chapter 2) using the ArF_2 or similar surface ablating laser to form a Fresnel prism in the anterior corneal stroma ΔPRK (Fig. 1-30, A). A shallow-depth Fresnel prism profile corneal ablation would not only allow higher corrections to be obtained without focal thinning but also permit more rapid healing and epithelialization with less persistent corneal haze.

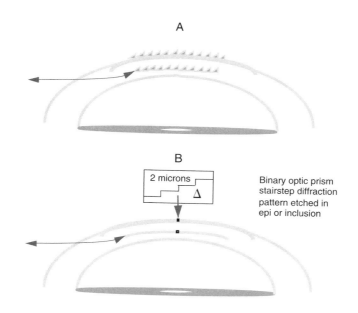

FIGURE 1-31. Schematic representation of "temporary" removable corneal prisms of uniform thickness. **A,** Fresnel principle applied to removable "epi" and intrastromal (scleral pocket) laser-cut prisms. **B,** Binary optics applied to "epi" and intrastromal (scleral pocket) laser-cut or ion beam–cut prisms. Microscopic pattern achieves smoother, thinner prism than Fresnel type.

Alternatively, applying the diffractive optics principle embodied in binary optics, the simplest form of which is the prism, a correction could be etched on the surface of Bowman's layer, further maintaining the structural integrity of the cornea (Fig. 1-30, B).

The "temporary," introduction of a prism correction can be accomplished also by surface fixation of corneal tissue, collagen, or an alloplastic material as an "epi" transplant in which a conventional, Fresnel, or binary optic prism has been formed or as an intrastromal corneal inclusion under a flap or pocket (Fig. 1-31, A and B). In theory these techniques would be reversible and could be used, following a trial with spectacle-mounted Fresnel or incorporated prisms, to determine the corneal prism power requirement as well as the efficacy of a permanent laser correction.

Prism automated lamellar photorefractive keratectomy (ΔALPK). Currently, automated lamellar keratoplasty (ALK) is being used in combination with intrastromal excimer laser ablation or laser-formed lenticle inclusions for correction of high myopic and hyperopic errors (ALPK). After a corneal flap has been cut with a microkeratome, a conventional prism, a Fresnel prism, or a binary optic prism could be formed within the stroma with a laser (ΔALPK) (Fig. 1-32 A and B) or with an ion beam, as currently done in optics.

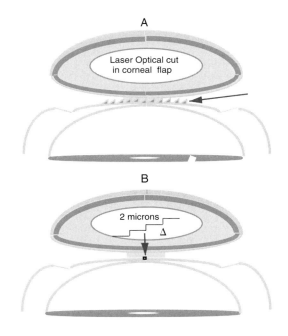

FIGURE 1-32. Schematic representations of laser-formed prisms in bed or cap of cornea following microkeratome partial (flap type) or complete corneal excisions (ΔALPK). **A,** Posterior stroma laser-cut Fresnel prism *(arrow)* combined with laser optical cut on corneal cap. **B,** Binary optic prism and optical stairstep diffraction pattern etched in posterior stroma *(arrow);* may be combined with optic cut of cap.

As with an "epi" prism, alloplastic or collagen prism inclusion could be tried before the permanent laser procedure.[24]

Another option would be to use, as originally proposed in 1986,[23] the picosecond (396 nm) laser to create intrastromal ablations without the necessity to disrupt Bowman's layer. This modality could be used to form an intracorneal Fresnel prism profile.

The mathematic calculation of the prism power induced by any of these modalities is dependent on the physical variables as well as the corneal optics.

REFERENCES

1. American Academy of Ophthalmology: Optics refraction and contact lenses, *Basic and Clinical Science Course, section 2*, San Francisco, 1989-1990, p 56.
2. Anschütz T: Theoretic possibilities of excimer laser: "Conoid Laser Keratectomy" to perform prismatic profile. *European Excimer Congress*, Strasbourg, 1990.
3. Azar DT: ArF excimer prismatic photokeratectomy in the treatment of consecutive small angle strabismus deviations, scientific poster 103, *Ophthalmology Suppl* 100:119, 1993.
4. Bicas HEA: Medidas angulares na rotina oftalmológica: variações do efeito de prismas em função de seus posicionamentos, *Rev Bras Oftalmol* 39:5-32, 1980.
5. Bicas HEA: Efeitos rotacionais mono e binoculares das associações de prismas, *Rev Bras Oftalmol* 39:33-45, 1980.
6. Duke-Elder S, Abrams D: *System of ophthalmology*, vol 5, *Ophthalmic optics and refraction*, St Louis, 1970, Mosby, pp 48-49.
7. Fannin TE, Grosvenor T: *Clinical optics*, Stoneham, Mass, 1987, Butterworth.
8. Fonda G: *Management of the patient with subnormal vision*, ed 2, St Louis, 1970, Mosby, pp 75-76, 84-91.
9. Fry GA: The major reference point in single vision lenses, *Am J Optom Arch Am Acad Optom* 24:1-7, 1947.
10. Hardy LH, Chace RR, Wheeler MC: Ophthalmic prisms, corrective and metric, *Arch Ophthalmol* 33:381-384, 1945.
11. Hardy LH: Clinical use of ophthalmic prisms (metric), *Arch Ophthalmol* 34:16-23, 1945.
12. Jackson E: The designation of prisms by the angular deviation they cause, instead of by the refracting angle, *Tr Int M Cong* 3 (sect 11): 785, 1887.
13. Moore S, Stockbridge L: Fresnel prisms in the management of combined horizontal and vertical strabismus, *Am Orthop J* 22:14-21, 1972.
14. Ogle KN: Distortion of the image by ophthalmic prisms, *AMA Arch Ophthalmol* 47:121-131, 1952.
15. Prentice CF: A metric system of numbering and measuring prisms, *Arch Ophthalmol* 19:64-75, 1890.
16. Putnam OA, Quereau JV: Precisional errors in measurement of squint and phoria, *Arch Ophthalmol* 34:7-15, 1945.
17. Reinecke RD, Simons K, Moss A, Morton G: An improved method of fitting resultant prism in treatment of two-axis strabismus, *Arch Ophthalmol* 95:1255-1257, 1977.
18. Repka MX, Arnoldi KA: Lateral incomitance in exotropia: fact or artifact? *J Pediatr Ophthalmol Strabismus* 28:125-130, 1991.
19. Risley SD: A new rotary prism, *Trans Am Ophthalmol Soc* 5:412-413, 1889.
20. Scattergood KD, Brown, MH, Guyton DL: Artifacts introduced by spectacle lenses in the measurement of strabismic deviations, *Am J Ophthalmol* 96:439-448, 1983.
21. Thompson JT, Guyton DL: Ophthalmic prisms: measurement errors and how to minimize them, *Ophthalmology* 90:204-210, 1983.

22. Thompson JT, Guyton DL: Ophthalmic prisms: deviant behavior at near, *Ophthalmology* 92:684-690, 1985.

23. Troutman RC, Véronneau-Troutman S, Jakobiec FA, Krebs W: A new laser for collagen wounding in corneal and strabismus surgery: a preliminary report, *Trans Am Ophthalmol Soc* 84:117-132, 1986.

24. Troutman RC: Personal communication, Nov 1993.

25. Veldkamp WB, McHugh TJ: Binary optics *Sci Am*, 92-97, May 1992.

26. Veldkamp WB: Binary optics. In *The McGraw-Hill encyclopedia of science and technology: 1990*, New York, 1989, McGraw-Hill, pp 39-42.

27. Véronneau-Troutman S: A new optical modality to overcome diplopia, *Trans Am Ophthalmol Soc* 77:181-190, 1979.

28. Welford WT: *Aberrations of the symmetrical optical system*, London, 1974, Academic Press, pp 73-110.

CHAPTER **2**

The Fresnel Principle

Fresnel Prisms

In 1821 Augustin Fresnel, a French physicist, introduced a new concept for the construction of converging lenses. At that time the large, high plus glass lenses, made primarily for use in lighthouses, were of considerable bulk and weight. In addition, the contemporary lens manufacturing technology made it difficult to construct them to the power necessary to concentrate the light sources then available toward the horizon, much of the light being lost peripherally. Because the weight of these large-diameter glass lenses required complex machinery to rotate the lens systems, a lighter lens promised to be particularly advantageous (Fig. 2-1, A).

Because the power of optical systems is essentially unaffected by changes in the thickness of the system elements or their separation, Fresnel reasoned that he could overcome both problems—inadequate power and excessive weight—by using a concentric set of prismatic rings, the face of each having the curvature and power of the lens element it replaced. The application of this idea resulted in more powerful converging lenses of much reduced thickness and weight (Fig. 2-1, B). These lenses, first used in lighthouses, still are used extensively in stage lighting, automobile headlights, and in a number of other optical applications.

As discussed in Chapter 1, the deviation of a prism equals the sum of the angles of incidence and emergence minus the apical angle. The thickness of the prism is not a factor in determining the prismatic deviation. The Fresnel principle is applied to flat-faced prisms with even a greater degree of accuracy than to lenses. The resulting reduction in weight is significantly greater in the Fresnel prism than it is in the Fresnel lens (Fig. 2-1, C). Each of the small, uniformly thick, juxtaposed prisms combined in a Fresnel membrane has the same deviating power as the thick, bulky, conventional prism it replaces.

Fresnel wafer prism. In 1965 Fletcher Woodward, an orthoptic technician, and Chester Rorie, an optician, collaborated in the design of the first ophthalmic Fresnel prism, which they called the "wafer prism."[26] This rigid, hard acrylic, Fresnel prism had one surface impressed with the prismatic grooves

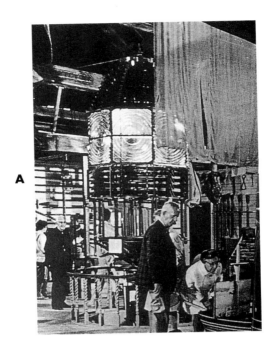

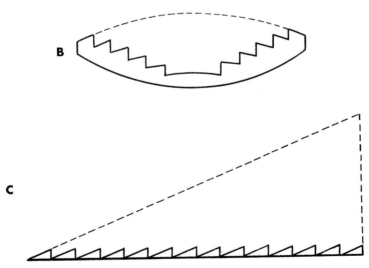

FIGURE 2-1. **A,** Sankaty lighthouse, 1849, from Henri Lepaute at the Whaling Museum in Nantucket, Mass. This lighthouse is one of the first structures to apply the Fresnel prism. Not until 1965 did Fresnel prisms become ophthalmic tools. **B,** Comparison of Fresnel plus lens with conventional biconvex lens. **C,** Comparison of Fresnel prism membrane to conventional prism.

(From Véronneau-Troutman S: *Trans Am Ophthalmol Soc* 76:610-653, 1978.)

and the other flat (Fig. 2-2, A, *left*). It was available first from 5Δ through 30Δ, and later up to and including 40Δ.

Although the wafer prism is of uniform thickness and is lighter in weight than the conventional prism, its striations are so apparent that in North America its application has been limited almost entirely to office use as a trial prism. Universal Optical Co. (Dallas, Texas) and Ellisor International, its most recent manufacturers, have discontinued production. The wafer prisms are made with larger diameters than the Fresnel trial set prisms (Fig. 2-2, A and B). Some examiners believe that the wafer prisms allow a better observation of the eyes in a wider excursion and are preferred as trial prisms.[19] In powers above 20Δ they reduce visual acuity less than the Fresnel membrane and are durable and reusable. For these reasons, despite their poor cosmetic appearance, wafer prisms were used for treatment.[16]

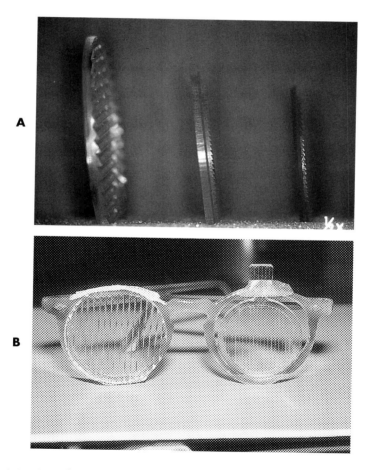

FIGURE 2-2. A, Side view of 30Δ Fresnel prisms. *Left,* wafer prism. *Center,* trial set prism. *Right,* membrane prism (Press-on). **B,** Front view comparing Fresnel hard plastic prisms, taped on carrier frame for demonstration purposes. *Right lens,* 35Δ wafer prism. *Left lens,* 35Δ trial set prism.

Fresnel membrane prism. In 1970 the Optical Sciences Group, Inc., introduced to the market the Fresnel membrane prism (Fig. 2-2, A, *right*) (Press-On).*[10] The flexible membrane material from which it is molded, optical-grade polyvinyl chloride (PVC), when wet conforms to a spectacle lens and adheres firmly once the water is absorbed. Like the wafer prism, one surface has the prismatic grooves, and the other is flat. Initially, the prism was made in powers ranging from 0.5Δ to 15Δ. In 1971 this range was extended to 30Δ, and in 1991 to 40Δ. Not only is this prism flexible, enabling it to conform to the base curve of the spectacle lens, but also it is thinner than the original wafer prism (see Fig. 2-2, A). Even at its maximum power its overall thickness does not exceed 2 mm (see Fig. 2-1). Comparatively, a prism ground in a 50-mm lens blank (index of refraction, 1.523) has an increasing thickness to its base of about 1 mm for each prism diopter of power.[7(p108)]

Fresnel Prism Trial Sets. In the mid-seventies the Optical Science Group, Inc., marketed a trial set of seven Fresnel prisms made of hard plastic material, which was thinner than the wafer prism but thicker than the membrane. This prism trial set, available in 12, 15, 20, 25, 30, 35, and 40Δ power, advantageously replaced the wafer prisms as a trial prism (Fig. 2-3).** Fortunately its production, discontinued for some years, has recently been resumed. In 1991 a second trial set with seven prisms of weaker powers became available (2 to 6, 8, and 10Δ), which can substitute for loose standard prisms when unavailable. In my opinion, *high-power Fresnel trial prisms (12Δ to 40Δ) are so*

---

*Press-On prisms are available in powers of 1 to 10, 12, 15, 20, 25, 30, 35, and 40Δ. They are currently manufactured by Health Care Specialties Division/3M, St. Paul, Minn 55144.

**3M presently manufactures the trial sets. They are distributed by The Fresnel Prism and Lens Co, Route #1 Siren, WI 54872. The author has no commercial interest in this product nor in any mentioned in this book.

FIGURE 2-3. The high-power Fresnel Prism Trial Set (12Δ to 40Δ). Individual powers of 12, 15, 20, 25, 30, 35, and 40Δ.

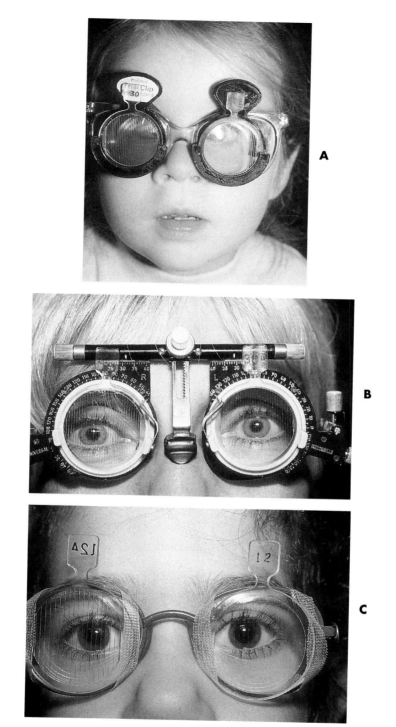

FIGURE 2-4. Fresnel prisms from trial set. **A,** Mounted over patient's spectacles in Halberg Clip-on lens cells (when in position taped to avoid rotation). **B,** In trial frame. **C,** Taped over patient's spectacles or light-weight plano frame for more comfortable wear during prolonged office trial.

superior to stan-dard prisms that they are *indispensable tools for the proper evaluation of ocular motility disturbances.* They allow a more accurate evaluation of large-angle deviations. They facilitate greatly the performance of any prism trial test in the office. Although larger diameter prisms would allow better observation of the eye, the present ones fit in commercially available trial frames and clip-ons (Fig. 2-4, A-C). However, sturdier handles and power markings that do not fade with time would be beneficial. Because a hard Fresnel prism reduces vision less and creates less distortion and aberration than a membrane of corresponding power,[23] the examiner can better predict the tolerance to a given membrane. The patient who finds a prism from the trial set unacceptable because of its side effects is obviously not a candidate for a Fresnel membrane.

Fresnel prism bars. The Optical Sciences Group, Inc., also introduced prism bars composed of Fresnel prisms to evaluate horizontal and vertical deviations (Fig. 2-5). With time and use they decrease vision substantially more than the Berens' prism bars, separate prisms, or Fresnel trial set prisms of corresponding powers; therefore the examiner must ensure that the patient actually sees the fixation symbol or light. Fresnel prism bars are wider than Berens' prism bars and therefore allow better observation of the eye; however, this can make it difficult to hold a horizontal and a vertical prism bar with the same hand so that the other hand is free to perform a cover test. Contrary to the conventional prism bar that has its prism powers printed on the side, the powers

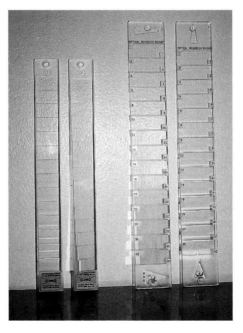

FIGURE 2-5. Horizontal and vertical Fresnel prism bars compared to conventional horizontal and vertical Berens' prism bars.

of the Fresnel prism bar are printed on the two faces, which facilitates the reading of the deviation by the examiner (Fig. 2-6 A and B).*

Effect of Fresnel prisms on visual acuity. *Wafer prisms, Fresnel trial set prisms,* and *conventional prisms induce a similar reduction* in visual acuity for corresponding powers. This reduction is one to two lines for prisms of powers greater than 12Δ and negligible below this power. Comparatively, the *Fresnel prism membrane reduces the visual acuity more in the higher powers.* A 30Δ membrane may decrease visual acuity from 0/20 to 20/40. Only with powers less than 12Δ is the visual acuity minimally affected.[23]

*Horizontal bars of 1, 2, 4, 6, 8, 10, 12, 14, 16, 18, 20, 25, 30, 35, and 40Δ. Vertical bars of 1 to 6, 8, 10, 12, 14, 16, 18, 20, and 25Δ. Production presently discontinued. Their reintroduction in a more scratch-resistant material would be welcome.

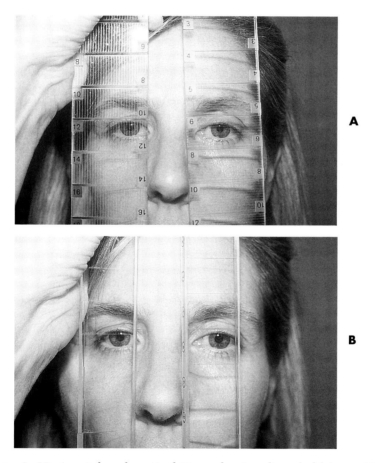

FIGURE 2-6. A, Horizontal and vertical Fresnel prism bars held in position for examination. Power easily read by examiner. **B,** Berens' type of bars being similarly used. Narrower width and clearer view facilitate examination but side power notations awkward to read.

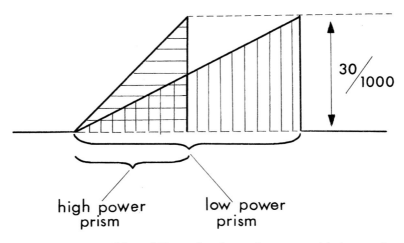

FIGURE 2-7. Groove widths of Fresnel prisms decrease with increasing power. At 30Δ, "critical point" smaller grooves create dispersion and reduce acuity.

(From Véronneau-Troutman S: *Trans Am Ophthalmol Soc* 76:610-653, 1978.)

This decrease in visual acuity is induced not because of the Fresnel principle, but rather from the material used and its thickness. The Fresnel trial set prism, made of solid plastic, has a total thickness of 3 mm, including the 1.5-mm plano backing. The membrane made of PVC, a flexible material, does not exceed 1 mm in thickness. The groove depth of a Fresnel membrane prism must not exceed 30/1000 inch.[1] To maintain this limit, decreasing groove widths must be used as the power of the prism increases until a "critical point" (Fig. 2-7) is reached and the width of the groove becomes too small, thereby resulting in dispersion and a precipitous reduction in visual acuity.

Although the flexibility of the membrane is a definite advantage, it makes it more susceptible to manufacturing variations. This can be well demonstrated by testing membranes of the same power from different lots. Especially for powers above 12Δ, the effect of the membrane on visual acuity should be determined for each patient and not assumed to only reduce vision by one or two lines, as often mentioned. With use and time the membrane becomes yellow, which further decreases vision. This change is accentuated in smokers, from eye cosmetics, and in some occupational environments. For cosmetic appearance and for optical reasons a patient should expect to have the membrane changed about every 6 months.

The reduction in vision is used to an advantage as a partial occluder in the treatment of amblyopia in the child. However, when the purpose is to develop binocularity, the decrease in vision and the increase of optical aberrations from higher powers limit the prescription of the membrane to powers less than 20Δ over either one or both eyes, using the best quality membranes presently available.

To reestablish fusion in the adult, the handicap of reduced acuity can be partially bypassed by combining prisms of different powers and materials. For example, a patient can be very annoyed with a 30Δ membrane over one eye or with a 15Δ membrane over each eye, despite the diplopia being corrected.

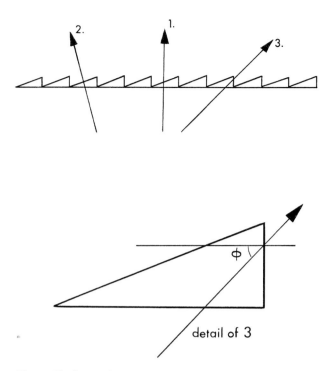

FIGURE 2-8. Effect of light striking a membrane prism. Rays *1* and *2* pass through the prism. Ray *3*, at the "critical angle," is deflected from the base away from the eye.

(From Véronneau-Troutman S: *Trans Am Ophthalmol Soc* 76:610-653, 1978.)

The same patient can tolerate well a 25Δ membrane over the nonpreferred eye and a 5Δ clear glass or plastic trial prism over the fellow eye. At first, incorporated prisms, 2.5Δ over each eye, are prescribed. Once the patient has the prismatic spectacle correction, the trial test is repeated, and then a 25Δ membrane, or less as required, is applied over the nonpreferred eye.

Effect of direction of Fresnel prisms on visual acuity. The difference in visual acuity between base-up and base-down, base-in and base-out is not significantly different for the Fresnel membrane prism and for the conventional prism as long as the patient looks straight ahead. *When the patient looks through the base of the prism, Fresnel membrane prisms of all powers and conventional prisms of 10Δ to 30Δ reduce visual acuity by one to two lines.*[23] There is no significant effect when looking toward the apex. The reduced acuity is due to the loss of light by reflectivity. The closer its incidence to the critical angle, the more reduced the visual acuity (Fig. 2-8). As a result the patient may adopt a face turn toward the base of the prism to direct the gaze away from it in an attempt to improve the vision.

Chromatic aberration and distortion. Optical PVC, from which the membrane prism is made, has been shown to increase chromatic dispersion and to produce a loss in contrast of objects when viewed through the prism.[9] Prisms from the Fresnel trial set, being of solid plastic material, have less optical aberration. With all types of Fresnel prisms, reflections from the prism facets induce a second image reflected toward the base of the prism. These reflections are especially annoying when looking toward a brightly lit object or a point source of light, as the prism facets produce a linear distortion like a Maddox rod.

Adams and co-workers[2] compared some of the distortions inherent in conventional prisms and in Fresnel membrane prisms with powers of 5Δ, 10Δ, and 15Δ when used in lenses with increasing base curves. These authors considered five of the distortions described by Ogle[13]: (1) horizontal magnification, (2) curvature of vertical lines, (3) asymmetric horizontal magnification, (4) vertical magnification, and (5) change in vertical magnification with horizontal viewing angle (see Distortion in Chapter 1). With a conventional prism, the overall magnification (horizontal and vertical) increases proportionally with increasing base curve. In contrast, the overall magnification induced by the membrane prism is minimal and relatively unaffected by a change in the base curve. With either type of prism, the other three distortions are reduced as the base curve of the lens is increased.

When a conventional prism is prescribed, in most instances, because of the magnification factor, the total power required should be divided equally between the two eyes, otherwise monocular magnification could induce aniseikonia. The minimal magnification induced by the Fresnel membrane prism is a definite advantage, allowing its prescription over only one eye.

Potential aberrations of membrane-spectacle lens combinations. The refractive index of the spectacle lens material to which the membrane prism is to be affixed is critical to its function.

Internal reflection. The polyvinyl chloride from which the membrane prism is formed has an index of refraction[12] of 1.525, approximating that of crown glass (1.518) (AO 1.523) or a CR-39 plastic (1.503) (AO 1.498) spectacle lens. Close index-matching of the lens and the membrane prism can eliminate problems of internal reflection that can occur if the index difference is greater than 1% or 2%.

Color separation. Theoretically, the degree of color separation that a prism introduces into a system is related directly to the difference in the refraction between the materials, each of which has a specific color wavelength. The standard measure of this relationship is the color number (V) defined as:

$$V = (nF - nC)/(nD - 1)$$

where:

nF = index of refraction for blue-green

nC = index of refraction for red

nD = index of refraction for yellow

For spectacle glass, V = 59.2, and for polyvinyl chloride, V = 59.3.[14] Therefore the addition of a flexible prismatic membrane to a crown glass or a CR-39 plastic spectacle lens should not induce chromatic effects beyond those of a conventional lens of a single material. Any lens made of a material of significantly different refractive index, for example, a flint glass lens (index of refraction = 1.621; V = 37.97), should not be used in conjunction with a PVC membrane prism because the combination would cause severe aberration problems.

Effect of Fresnel prisms on binocularity. In a previous study[23] when normal subjects were fitted with membranes of 20Δ, 25Δ, and 30Δ, fusion and stereopsis were disrupted to such an extent that the use of membranes of these powers to restore or improve binocular vision was seriously questioned. Also, the disruption of fusion and stereopsis was abrupt and severe and did not parallel the decrease in visual acuity. Membranes of lesser powers affected the binocular state less deeply and less frequently.

In the same study it was found that hard Fresnel prisms (from the trial set) of corresponding powers did not disrupt fusion and stereopsis. For powers *above 20Δ the hard plastic Fresnel prism from the trial set will determine more accurately the sensory state* of a patient. Therefore in these powers, the hard Fresnel prism *should be used instead of the membrane for a PAT.* If such prisms were made more acceptable cosmetically and fitted to a special round adjustable frame, as that designed by Bérard in the late 1960s for mounting standard prisms up to 10Δ or wafer prisms in higher powers over an optical correction (Fig. 2-9), prismotherapy to establish binocularity in large-angle strabismus may be reconsidered.

Technique of cutting and applying the Fresnel prism membrane. A great advantage of the membrane is that it can be readily applied in the examiner's office at the completion of the examination and prism trial. It can be cut to fit the patient's glasses according to a direct tracing from a pantograph that has been designed for this purpose (Fig. 2-10, A and B). Alternatively, a template,

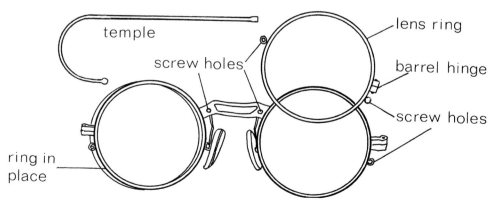

FIGURE 2-9. A special frames designed by Bérard to accept standard or Fresnel wafer prisms over an optical correction.

(From Véronneau-Troutman S: *Trans Am Ophthalmol Soc* 76:610-653, 1978.)

which can be retained for future use, can be made by pressing a paper firmly to the internal aspect of the patient's spectacle frame or outer edge of a lens. The pattern thus obtained is cut about 1 mm inside the tracing to fit within the frame periphery to facilitate the application and maintenance of the membrane, which is cut to fit the pattern and applied to the spectacle lens (Figs. 2-10, B, and 2-11, A-C). A membrane fitted too closely to or touching the rim of the frame will roll at the edge and trap air bubbles, thereby loosening its attachment to the lens surface. To apply the membrane it is *totally submerged* together with the spectacles in lukewarm water. Then the smooth back surface of the prism membrane is pressed onto the posterior surface of the spectacle lens (Fig. 2-12, A and B). All air bubbles must be expressed as it is positioned. The submersion technique achieves a more uniform, lasting result than the practice of using a wetting solution to affix the membrane to the lens surface.

After a few hours the membrane adheres securely to the lens as the water between the apposed, smooth surfaces is absorbed by the PVC. The patient is

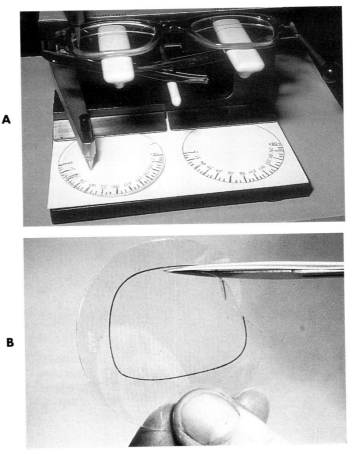

FIGURE 2-10. A, Pantograph for outlining pattern for fitting Fresnel prism membrane to spectacle lens. Protractor scales fitted by the author facilitate membrane orientation. **B,** Membrane pattern must be cut 1 mm inside the tracing.

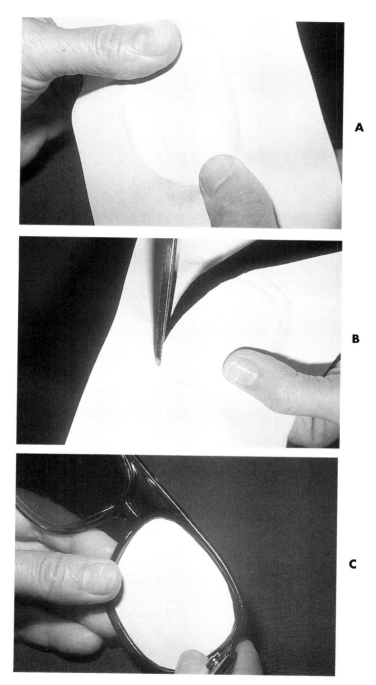

FIGURE 2-11. A, Technique for obtaining a pattern by pressing paper against the patient's spectacle frame periphery. **B,** Template being cut 1 mm inside the imprint pattern. **C,** Verifying fit before using paper template to cut the membrane.

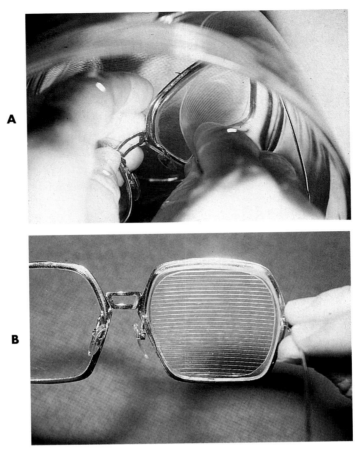

FIGURE 2-12. A, Immersion technique for applying smooth surface of membrane to posterior surface of spectacle lens. **B,** Properly cut and applied membrane (1 mm inside frame, no bubbles or raised edges).

told not to clean the glasses before the next day. Even when a good seal has been created the membrane can be easily removed by lifting one edge with the fingernail. These prism membranes are not glued or "pasted on".

For patients who do not wear an optical correction the membrane can be applied initially to a pair of lightly tinted sunglasses, which are readily available with little expense. The tint will also make the striations less noticeable to the observer and improve the appearance.

Cleaning and reapplication of the membrane. The patient or the parents are instructed on how to clean the membrane properly. The surface charge of PVC attracts dirt and dust more than does the glass or hard acrylic used in the manufacturing of other prisms. PVC material is sensitive to certain products. Cleaning solutions containing acetone or alcohol will damage the membrane and should not be used. To clean the spectacles, they are agitated in a basin of lukewarm water to which a few drops of a mild detergent have been added, then rinsed, and dried with a lint-free material.

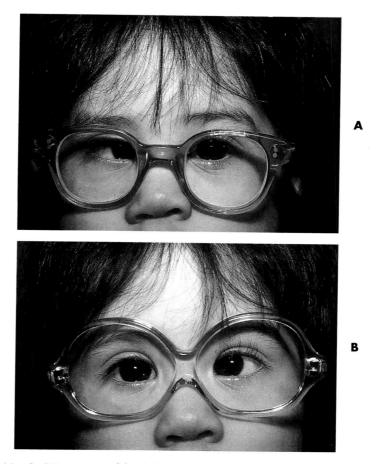

FIGURE 2-13. A, Miniaturized high-bridge adult frame. Unsuitable for child cosmetically and functionally. **B,** Same child wearing properly designed and fitted frame.

The patient or the parents are also instructed on how to reapply the membrane if it is accidentally detached. Unless the frame is round or perfectly oval, the contour of the frame can orient the reapplication underwater of the smooth part of the membrane to the posterior surface of the lens. The correct orientation should be verified by the examiner as soon as possible.

With proper maintenance the membrane will be functional for 6 months, exceptionally up to 1 year or more, at which time it has to be changed because of yellowing, decreased visual acuity, or unsightly appearance.

Frame Selection

Until recently, most manufacturers have produced unsuitable, often unsightly miniaturized adult frames (Fig. 2-13). Now optical shops exist that specialize only in children's frames.

It is important that the prismatic as well as the optical correction be mounted in a proper frame. However, it has been more difficult for a baby or-

FIGURE 2-14. Specially designed children's frames with bridge of correct width and height for young age group. Child correctly fitted with **A,** low, narrow bridge; **B,** low, wider nose bridge; **C,** higher bridge, eyes well centered in frame; **D,** economical solution, unsuitable frame reshaped by resourceful optician to obtain similar fit to **C.**

a young child to be well fitted because few companies have frames designed specifically for the facial configuration of a child (Fig. 2-14, A-D).

Nevertheless, a child's frame should, at a minimum, have the following characteristics:

1. The frame should have a low-fitting bridge that prevents it from sliding down the lower anatomic nose bridge of the child. It should distribute the weight of the eyeglasses evenly on the child's soft cartilaginous nose tissues without being too wide; otherwise, the frames will rock on the bridge. The eyes should be centered to the lenses. A discrete, replaceable silicone cushion can be added to keep the glasses more firmly in place but only if applied to a well-designed frame (Fig. 2-15).

2. Temples with spring hinges better endure the active lifestyles of a child, and these should be long enough to wrap around the ears. They should not pull the ears forward nor press on the mastoid areas.

One frame* that fits all of the preceding criteria has been designed especially for the young child. The temples of this frame are bent to conform to the back of the ear at three designated points. They also have a hole in their tips to allow an elastic or a ribbon to be attached to hold the frame from behind. This may be contraindicated in children under the age of 3 for fear that a child may choke on it; however, this should not be a concern when the child is under supervision.

For children, plastic frames are preferred to metal frames because they are lighter and safer. Elliptical-shaped, aviator-style frames that allow the child to look between the frame and the nose should be discouraged. Often poor fitting is blamed on the shape of the child's nose bridge, and awkward remedies, such as an unsightly molded silicone nosepiece, have been proposed[20]

---

*Manufactured by Les Frères Lissac Opticiens of which Meyrowitz Inc. in the United States, is a subsidiary. Jean-Pierre Bonnac, Technical Director, has been an innovator and for over 25 years, has designed frames for children as young as 3 months of age.[3,4]

FIGURE 2-15. Child's frame fitted with discrete, adjustable silicone cushion to provide additional support.

when in fact the frame design is the problem. Unless there is a severe patho-logic deformity of the nose bridge or the ears, *all patients, despite age or race, can be fitted with an appropriate frame that is also cosmetically acceptable.* Adult styles and colors may attract the older child, but the parent should be watch-ful that the final frame selected fits well or the uncomfortable but stylish glasses are not functional. Glasses are always heavier when lenses, especially with prisms incorporated, are mounted.

Single-vision lenses and bifocals with disparate power prisms. Fresnel membranes can be applied to all types of bifocals, including progressive, for both the child and the adult. When the patient has diplopia except when read-ing, the membrane is cut off over the bifocal (Fig. 2-16, A). Conversely, when the patient has diplopia only when reading, the membrane may be cut to cover only the bifocal. If different orientations or powers are needed at near and at

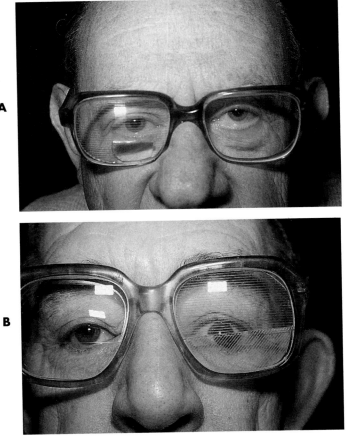

FIGURE 2-16. A, Fresnel membrane for vertical strabismus with cut-off to ex-clude bifocal segment. **B,** Fresnel membranes of different powers and orientations (top vertical prism, lower oblique prism) fitted to upper and lower sectors of a bifo-cal lens.

distance, the lower membrane may be more functional and better tolerated if it is cut to cover the entire inferior half of the lens including the bifocal (Fig. 2-16, B). However, a trial with a single membrane should be given first if the difference between the near and distance prism power is not significant.

Prisms for Field Defects

To reorient the field of vision vertically (e.g., in a recumbent patient) or to displace the residual field in conditions such as retinitis pigmentosa and hemianopia, a mirror or a mirror combined with cemented prisms has been strategically positioned over a segment of the spectacle lenses.[6] Fresnel membranes alleviate some of the disadvantages of these systems. Not only are they less expensive, less fragile, and lighter, but also they allow larger corrections and facilitate a change in the separation between the prisms in case the range of vision to either side of the displaced field is insufficient.[24] Unlike conventional prisms, they are not affected by the base curve of the carrier lens.

However, as with other systems, the patient has to be highly motivated and able to adapt. The patient has to be stationary when changing fixation and has to tolerate the "jump," which occurs when looking from a carrier lens to a zone with a prism.

Tubular field. In the case of a tubular field, a base-in prism is applied nasally and a base-out prism is applied temporally to the carrier lens to bring peripheral objects into the patient's field of vision. Patients with 5° to 15° of field and visual acuity of 20/100 or better are the best candidates. Patients with smaller fields do poorly because they have acquired a compensatory "scanning" technique and have become accustomed to it.[15]

Patients with less than 5° of field with no peripheral island of vision may be good candidates if they have adopted a fixed gaze.[8] Patients with a 20° field or larger do better with mobility training and scanning techniques. Acclimation to the "jack in the box" phenomenon when switching to or from the prism and the apparent increase in the distance of objects as they are viewed through the prism are also important in the training and acceptance of this therapy.

Weiss[24] has devised a technique to determine the sectors of the lens to be covered with the prism. With the head in the primary position the patient fixates at distance with one eye occluded. Precut paper strips of increasing width are placed vertically in front of the spectacle lens of the fixing eye until the residual temporal and nasal fields of vision are just blocked out. Without any change in the eye or head position the process is repeated for the fellow eye. When the optimal position of the occluding strips of paper is determined, they are taped to the front of the spectacle carrier lenses over each eye (Fig. 2-17, A). While continuing to caution against any eye movement, the examiner then asks the patient to keep both eyes open and move the head to the left and then toward the right to recheck the placement of the strips. The strips have to be positioned so that each eye begins to see past the occluding strips *at the same time*. Then the apical edge of a 15Δ base-out Fresnel membrane is ap-

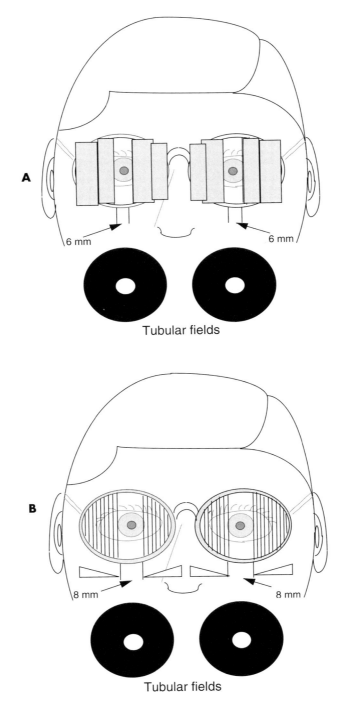

FIGURE 2-17. Prism membranes to extend the temporal and nasal fields of gaze to compensate tubular fields. **A,** Paper strips taped in place after optimal positioning. **B,** With strips removed and four 15Δ Fresnel membranes in place 1 mm distal to the central vertical edges of the strips, temporal base-out, nasal base-in.

plied to the back of the lens 1 mm external to the edge of the temporal paper strip, and a 15Δ base-in membrane is applied 1 mm internal to the edge of the nasal paper strip, parallel to the edge of the first membrane (Fig. 2-17, B). The lower sector of each nasal membrane can be cut off to facilitate convergence at near or in the presbyopic patient to coincide with the upper edges of the bifocals for reading. Recently, Weiss has reported an unusual case with small central fields and several peripheral islands of vision better helped when the edges of the temporal and nasal prisms were abutted along the visual axes.[25]

Opinions have differed regarding the individual prism powers (30Δ or 20Δ in place of 15Δ advised for both temporal and nasal prisms) and the sequence of fitting. The temporal prisms are sometimes fitted first, and only when the patient has become acclimated to these are the nasal prisms placed. Also reported is that the space between the temporal and nasal prisms may need to be increased as the patient adjusts to the visual aid. Base-down prisms have been added below the horizontal ones to increase the inferior vertical field but are too disorienting for constant wear.[8]

Homonymous hemianopia. In homonymous hemianopia the vertical edge of a Fresnel prism of appropriate power is applied to each lens with its base toward the affected field (Fig. 2-18), about 2 mm distal to the pupillary zone of the lens to avoid interference with macular vision. In a randomly assigned study involving 39 patients with stroke and homonymous hemianopia or unilateral visual neglect, Rossi et. al.[18] found that the subjects fitted with conju-

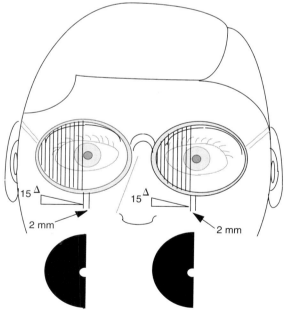

Right homonymous hemianopia

FIGURE 2-18. Position of two 15Δ Fresnel membrane prisms, bases in the direction of the defect, to extend the right field of gaze to compensate for a right homonymous hemianopia.

gate 15Δ Fresnel membranes outperformed the control group for all visual perception tasks, irrespective of the degree of observer involvement. Fifteen-diopter Fresnel membranes were selected because previous pilot studies had indicated that this was the greatest strength that could be easily tolerated. Ambulation and transfer safety were assessed by the patient's physical therapist. Patients were allowed to resume previous activities only when they demonstrated adequate visual adaptation and safety awareness. Most subjects tolerated the prisms well after 1 or 2 days.

Conjugate prisms can also be prescribed for a patient with restricted head movement to expand the horizontal or vertical gaze.[5;11] The base of the prism is oriented toward the restricted field in the same position as for a visual field defect.

Macular dysfunction. Prisms have also been advocated to improve the performance of patients with macular dysfunction. The better seeing eye is tested first. The patient is instructed to turn the head to place the eyes in the direction that gives the clearest vision. Then the examiner places an 8Δ trial prism over the eye with the better vision with the prism base in the direction of the head turn. If vision does not improve as the head is slowly turned, the patient is asked to rotate the prism until the best vision is obtained. The same procedure is repeated over the eye with the poorer vision. The prism correction is refined binocularly by having the patient rotate the prism over the fellow eye with the lesser vision to the position of maximum comfort with binocular single vision. Because of the mixed success reported clinically for this method, called "prisms relocation" or "scanning," Rosenberg et. al.[17] did a randomized study of 30 patients. After an adequate examination and trial, they concluded that the method was valid as an optical approach to the symptomatic functional improvement of visual performance in macular dysfunction.

Binary Optics

An advanced implementation of the Fresnel principle, binary optics[21;22] (a diffractive optics technology), has only recently become commercially feasible through the use of computer-assisted design (CAD) and the very large integrated scale (VLIS) circuit etching technology pioneered by the computer chip industry. *The two-level surface produced by a single etching step is the origin of the term binary optics.* Because binary optical devices control light by diffraction rather than by refraction, an optical device based on binary optics breaks up the wave front of incoming light at each point of the lens surface and reconstitutes it as a wave traveling in the desired direction to the point of focus. *In spherical optics it is already used as an advanced implementation of the Fresnel zone pattern, which uses circular phase gratings to focus light into a corrected spot.*

Binary optics are created by computer-generated, three-dimensional, blaze-like surface relief patterns, which are ion beam etched into an optical surface. Multiple steps (up to 16) are required for a diffraction efficiency of better than 99%. In addition, these surfaces can be designed to achromatize optical materials (Fig. 2-19) and for antireflective coatings.

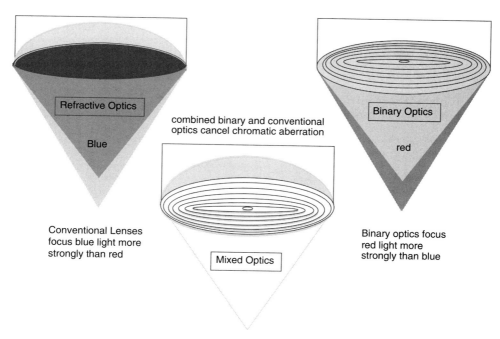

Refractive Optics

Blue

combined binary and conventional optics cancel chromatic aberration

Binary Optics

red

Conventional Lenses focus blue light more strongly than red

Mixed Optics

Binary optics focus red light more strongly than blue

FIGURE 2-19. Chromatic aberration neutralized by combining refractive and binary optics.

Binary optical devices have been designed to control amplitude, phase, and polarization of monochromatic or laser light for applications such as prisms, lenses (e.g., contact lenses and bifocal intraocular lenses), beam splitters, beam multiplexers, and filters. *The simplest binary optical structure is said to be the prism,* which appears under the microscope as a series of tiny staircases (see Figs. 1-30 and 1-31). When light falls on the surface of a binary optical prism, the wave is broken up into secondary wave fronts. Each wave front is delayed one full wavelength in proportion to the thickness of the staircase at that point. As the wave fronts interfere, the light is bent. Instead of the few millimeters of glass or plastic required to bend light in a Fresnel or conventional prism, a binary optical prism requires as little as 2 microns of material. For clinical purposes, using a combination of materials, binary optical prisms for spectacles or contact lenses could be designed to be thin at all powers and to have no apparent striations and little or no distortion. Unlike current photorefractive laser keratectomy, binary optical lenses or prisms would require only a few microns of tissue to form and could be etched into the anterior surface of Bowman's layer or intrastromally, preserving the corneal architecture (see Laser-Induced Corneal Prism in Chapter 1). In addition, chromatic aberration and reflection, which further reduce visual acuity with conventional and Fresnel prisms, could be corrected with binary optical prisms. These ideal prisms may be in the immediate future.

Current research in binary optics centers on extending the technology to broad waveband and microoptic applications. So called "fly's eye" arrays of microoptic binary optics may eventually be used to extend the perceived environment in patients with severely restricted visual fields as described with Fresnel prisms (see Figs. 2-17 and 2-18).

REFERENCES

1. According to letter from Torrance, AL, President, *Optical Sciences Group,* San Rafael, CA Jan 1977.
2. Adams AJ, Kapash RJ, Barkan E: Visual performance and optical properties of Fresnel membrane prisms, *Am J Optom Arch Am Acad Optom* 48:289-297, 1971.
3. Bonnac JP: La lunetterie du jeune enfant: Considérations physiologiques et morphologiques, *Bull Soc Sci Correction Oculaire* 1:85-94, 1985.
4. Bonnac JP: Equipements optiques et strabismes accommodatifs, *Bull Soc Sci Correction Oculaire* 1:107-112, 1985.
5. Cohen MB, Miles A: Fresnel prisms aid the arthritic, Case Reports, *Rev Optom* 119:70-74, 1982.
6. Duke-Elder S, Abrams D: *System of ophthalmology,* vol. 5, *Ophthalmic optics and refraction,* St Louis, 1970, Mosby, pp 703-712.
7. Fannin TE, Grosvenor T: *Clinical optics,* Stoneham, Mass, 1987, Butterworths, p 108.
8. Ferraro J, Jose R, Olsen L: Fresnel prisms as a treatment option for retinitis pigmentosa, *Tex Optom* 38:18-20, 1982.
9. Flom M, Adams JA: Fresnel optics. In Duane DT, editor: *Clinical ophthalmology,* vol 1, New York, 1976, Harper & Row, pp 1-15.
10. Jampolsky A, Flom M, Thorson JC: Membrane Fresnel prisms: a new therapeutic device. In Fells P, editor: *The First Congress of the International Strabismological Association, Acapulco, Mexico, 1970,* London, 1971, Henry Kimpton, pp 183-193.
11. Lalle P: Case Report: Unique case of vertical field expansion with Fresnel prism, *N Engl J Optom* 37:14-18, 1985.
12. Levi L.: *Applied optics,* vol 1, New York, 1968, John Wiley & Sons, p 584.
13. Ogle KN: Distortion of the image by ophthalmic prisms, *Arch Ophthalmol* 47:121-131, 1952.
14. Ogorkiewicy, RM: *Engineering properties of thermoplastics,* New York, 1970, John Wiley & Sons, p 251.
15. Perlin RR, Dziadul J: Fresnel prisms for field enhancement in patients with constricted or hemianopic visual fields, *J Am Optom Assoc* 62:58-64, 1991.
16. Pigassou-Albouy R: Traitement du strabisme par les prismes, *Doc Ophthalmol* 60:45-69, 1985.
17. Rosenberg R, Faye E, Fischer M, Budick D: Role of prism relocation in improving visual performance of patients with macular dysfunction, *Optom Vis Sci* 66:747-750, 1989.
18. Rossi PW, Kheyfets S, Reding MJ: Fresnel prisms improve visual perception in stroke patients with homonymous hemianopia or unilateral visual neglect, *Neurology* 40:1597-1599, 1990.
19. Shippman S, Weintraub D, Cohen KR, Weseley AC: Prisms in the preoperative diagnosis of intermittent exotropia, *Am Orthop J* 38:101-106, 1988.
20. Tate GW, Stenstrom WJ, Kulick C: A molded silicone nosepiece for children's eyeglasses. In Notes, cases, instruments, *Am J Ophthalmol* 78:726, 1974.
21. Veldkamp WB, McHugh TJ: Binary optics, *Sci Am,* 92-97, May 1992.
22. Veldkamp WB: Binary optics, In *The McGraw-Hill Encyclopedia of Science and Technology: 1990,* New York, 1989, McGraw-Hill, pp 39-42.
23. Véronneau-Troutman S: Fresnel prisms and their effects on visual acuity and binocularity, *Trans Am Ophthalmol Soc* 76:610-653, 1978.
24. Weiss NJ: An application of cemented prisms with severe field loss,

Am J Optom Arch Am Acad Optom 49:261-264, 1972.

26. Weiss NJ: An unusual application of prisms for field enhancement, *J Am Optom Assoc* 61:291-293, 1990.

27. Woodward F: Unlikely looking prisms. Presented to the American Association of Certified Orthoptists at the Seventieth Annual Session of the American Academy of Ophthalmology and Otolaryngology, Chicago, 1965.

CHAPTER 3

Tests Using Prisms

Ophthalmic prisms are produced commercially in several different forms and are calibrated differently according to the optical material used in their manufacture. Most trial lens sets come supplied with round, loose prisms, generally made of glass, of powers up to 20Δ. They are usually *calibrated to be positioned in the trial frame in the Prentice position (one face perpendicular to the line of sight* (see Fig. 1-5). Individual square prisms in powers from 1Δ to 50Δ are available in boxed sets for which light plastic such as methylmethacrylate *(calibrated in the minimum deviation position)* has replaced the heavier but more scratch-resistant glass *(calibrated in the Prentice position).* Prisms also are made joined successively in a horizontal or vertical orientation in bars in which the powers progress stepwise to reach 25Δ for the bars of vertical prisms and 40Δ and 45Δ for the bars of horizontal prisms. For powers below 20Δ the progressive steps vary from 1 to 2 diopters, and above 20Δ they increase in 5Δ steps. Prescription lenses, as well as some prism sets and bars, are available in harder plastic materials such as polycarbonate, which is lighter and more scratch resistant than methylmethacrylate. *Prisms made of plastic are usually calibrated for minimum deviation. However, for practical purposes, during use they are held in the frontal plane position (i.e. with their posterior surface perpendicular to the direction of the fixation target) to approximate the position of minimum deviation.* (See Figs. 1-5 and 1-6 and Tables 1-1 and 1-2.)

The rotary prism described by Risley[31] in 1889 is still in common use. It consists of two 15Δ glass prisms mounted back to back that rotate in opposite directions within a calibrated frame. As they are rotated from zero, when their bases are opposite to each other, their power increases progressively to a maximum power of 30Δ when their bases coincide. Most phoropters come equipped with two Risley prism units, the buyer having a choice of their effective powers. By flipping them over 90° a base-in or base-out effect (see Fig. 1-8), or alternatively a base-up or base-down effect, can be achieved (see Fig. 1-9). They provide a smooth, progressive increase or decrease in prism power, rather than stepwise, that is of particular advantage in the study of vergences. Because these prisms are mounted aligned with Maddox rods, also rotating 90°, in most phoropters their combined use is facilitated.

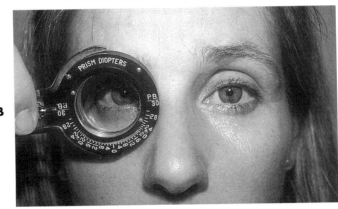

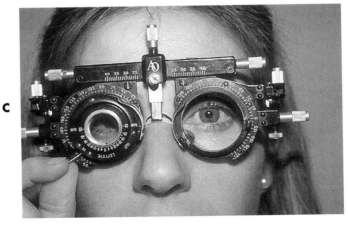

FIGURE 3-1. **A,** Risley prisms for use as independent instruments, to be hand held *(left)* or clipped in a trial frame *(right)*, in position for measuring vertical deviations or vergences. **B** and **C,** In position for measuring horizontal deviations or vergences. In illustrations the prisms are oriented base-out.

Originally described by Risley as a single instrument, the rotary prism is still available in the same forms; that is, in a hand-held unit or in one that can be clipped in a trial frame (Fig. 3-1). Unlike the phoropter mounted units, both can be oriented for an oblique prism effect.

Effects of Prisms on Movement of the Eyes

Monocular prismatic effects. The image of an object is displaced toward the base of the prism. However, for an observer viewing an object through the prism it will appear displaced toward its apex (see Fig. 1-2). To look at the object the observer's eye must move through an angle equal to the deviation of the prism. A base-out prism causes the eye to move inward (adduction), and a base-in prism causes the eye to move outward (abduction). When the prism is base-down the eye moves up (supraduction), and when the prism is base-up the eye moves down (infraduction). The fellow eye, behind an oc-cluder, is also seen to make a duction movement in the direction of the apex of the prism. This response is not affected by the presence of the occluder (Fig. 3-2).

Binocular prismatic effects. In an individual with normal binocular vision, when a prism is introduced before one eye with both eyes uncovered, the ini-tial corrective movement of both eyes is in the direction of its apex. When the deviation induced is within the fusional range, a movement of redress of the eye without the prism follows immediately. Using this response, deviations are neutralized and *vergences* are studied (Fig. 3-3).

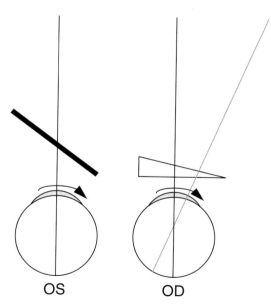

OS   OD

FIGURE 3-2. Monocular prismatic effects. With a base-in prism over one eye and the fellow eye occluded, so it can be observed by the examiner, both eyes make a duction movement toward the apex of the prism (conjugate movement).

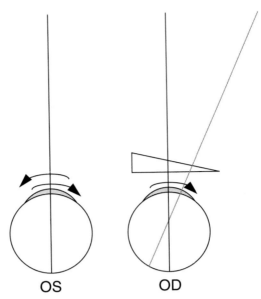

OS   OD

FIGURE 3-3. Binocular prismatic effects. With both eyes open, if the base-in prism power is within fusional divergence, the adducted left eye will abduct and single vision will be maintained. A disjunctive vergence movement has taken place.

However, with a base-in prism over one eye and a base-out prism of equal power over the fellow eye, each eye moves equally toward the apex of the respective prism. As the lines of sight in the displaced gaze position remain parallel there will be no attempt to converge and no movement of redress (Fig. 3-4). Such identically oriented prisms, placed over both eyes, called *conjugate prisms* (homonymous base or apex, or yoked), are used for the evaluation and treatment of abnormal head posture and field deficiencies. The bases of the prisms are always placed in the direction of the face turn, the chin position, or the tilt to be corrected (See Conjugate Prisms in Chapter 6), or from the nonseeing to the seeing field (See Chapter 2).

Prism and alternate cover test. In a cooperative patient with foveal fixation in both eyes, the use of a prism with an alternate cover test is the method preferred to objectively evaluate the angle of deviation. *All through the test, care must be taken that the head of the patient remains straight, that the fixation target is seen, and that accommodation is elicited.* With a known visual acuity an optotype slightly larger than the one resolved with the eye with lesser vision is chosen. It is also prudent to ensure that the patient sees the optotype through the often well used, scratched plastic prisms or through prisms of higher powers either of which can decrease vision and make fixation unreliable.

With children it is important not only that they visualize the fixation target but also that they maintain strict attention. A "wiggling" picture affixed to a tongue depressor that the child holds against the examiner's nose is a use-

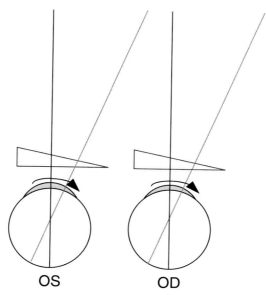

OS   OD

FIGURE 3-4. Conjugate prismatic effects. With a base-in prism over one eye and a base-out prism of equal power over the fellow eye, the lines of sight remain parallel, deviated toward the apex of the prisms.

ful and successful technique. It not only ensures that accommodation is being used but also keeps the examiner's hands free to manipulate the cover and the prisms. At distance, in a child of age 6 months to 4 years, a remote-controlled, lighted toy will both attract attention and calm an excited child. I have found such toys more efficient than tapes of television shows or movies that are more familiar to a child and thus less likely to hold their interest.

Although fixation targets that elicit accommodation are recommended, it is useful to compare the deviation elicited without accommodation when a patient is fixating on a point of light. A discrepancy will identify an accommodative component, easily demonstrated in accommodative esotropia but often overlooked in intermittent exotropia. For example, a patient with intermittent exotropia may have a manifest deviation when looking at a light but is phoric when looking at an accommodative target. The same patient may have better visual acuity with each eye separately than with both eyes open. This patient attempts to control the deviation by overaccommodation, which induces a pseudomyopia.

Technique. For the patient with an optical correction and maintaining accurate fixation, a prism of approximate power required to compensate the deviation is placed over the nonpreferred eye with its apex in the direction of the deviation. As the eyes are alternately covered and uncovered, the power of the prism is increased until the deviation is visibly converted to a movement in the opposite direction. The prism power is then decreased until the turn is abolished. This end point represents the objective horizontal or vertical angle of the deviation.

It is important that the eye not be allowed to recover from the dissociation induced by the cover test. It is crucial also to maintain the cover over one eye when changing the prism power. Using prism bars facilitates the technique. When evaluating combined deviations, a vertical bar and a horizontal bar held together, back to back, over the same eye or one bar over each eye held with the same hand gives the examiner a free hand to perform the alternate cover test (see Fig. 2-6, A and B). In combined deviations the greater deviation is corrected first. The neutralizing prism is left in place as the lesser vertical or horizontal deviation is corrected in turn. The effects of neutralizing the deviations individually is quickly ascertained. In incomitant deviations the measurements are repeated with the fellow eye taking fixation.

To measure large-angle deviations, *Fresnel trial set prisms have a definite advantage* of allowing a better observation of the eyes while reducing vision less (see Fig. 2-3). Because they are calibrated for the Prentice position, they can be best positioned perpendicular to the visual axis with the smooth surface placed toward the eyes for use in a trial frame or over the patient's spectacles in a pair of Halberg Clip-On lens cells (See Fig. 2-4, A and B).

In a patient with nystagmus the alternate cover and prism test (ACPT), although somewhat difficult, can be performed more readily with Fresnel trial set prisms. Nevertheless, in some patients a subjective test such as the Maddox rod may be easier to interpret and a very large angle may be evaluated better by the Krimsky test.

Alternate cover and prism measurements are best done at 33 cm and at 6 m, in straight-ahead gaze, and in four diagnostic positions: horizontally, to the right and to the left; and vertically, upward and downward.

The test is most reproducibly performed at distance by positioning the fixation targets at predetermined points in relation to the patient's head. However, at near, unless the examiner uses a deviometer type of instrument[28(pp175-176)], it is more difficult to always do the measurements at exactly 33 cm, 35° from the primary position. The presence of a V or A pattern is confirmed by taking measurements at distance before and after the prism adaptation test (PAT) (see Chapter 4).

When the ACPT reveals a vertical deviation, then measurements are taken in the oblique positions of gaze (up and to the right; up and to the left; down and to the right; and down and to the left) at about 35° peripheral to the primary position. When the ACPT is used without prisms, solely for diagnostic purposes, it can be performed more eccentrically.

A scale of −1 to −4 and +1 to +4 has been adopted by some authors to quantitate muscle actions. However, in patients with central fixation in both eyes, this technique for estimating muscle function cannot compare in accuracy to measurements obtained with the ACPT done in the nine diagnostic positions of gaze.

Measurement of incomitant deviations. When measuring a comitant deviation it is usually not important to record the fixating eye or the eye in front of which the prism is placed. However, when measuring primary and secondary deviations in an incomitant muscle disorder such as neuroparalytic strabismus, it is important always to indicate the eye with the prism. In such cases the primary deviation is produced when the nonparetic eye fixates. Conversely, when the paretic eye fixates, a secondary deviation results, which is larger than the primary deviation as a consequence of Hering's law of equal innervation to yoke muscles. Therefore the primary deviation is measured with the prism in front of the paretic eye. The secondary deviation is then measured by placing the prism in front of the sound eye while the patient fixates with the paretic eye. If there is a strong preference for one eye or a mechanical factor, the blur induced by the prismatic correction over the sound eye may not be sufficient to allow fixation to be maintained by the fellow eye. The examiner, while cautioning the patient to fixate, briefly covers the sound eye to verify that fixation has indeed been taken with the fellow eye.

Hess-Lancaster red and green streak test. This test, as described originally by Hess and later modified by Lancaster,[19(pp417-425);26] is based on the presentation of a different stimulus to each fovea (Fig. 3-5). Diplopia is created by stimulation on noncorresponding points in contradistinction to diplopia tests where a single stimulus is presented.

The test is performed in the following manner. After a trial with the streaks alone to check the accuracy of the response, the patient is equipped with colored goggles—red over the right eye and green over the left eye—and given the flashlight, which projects the bright red streak of light while the examiner projects the green one. The goggle lens colors are saturated so that each streak is seen only by the eye that wears the lens of the same color. In a dimly lit room, seated facing a calibrated grid screen *with the head held straight and immobile*, the patient is asked to superimpose the red streak onto the green streak, which the examiner projects on the screen in the diagnostic positions. The results obtained for the right eye, while the left eye is fixating, are manually recorded to a calibrated chart matching the grid-screen parameters. I perform the test at 1 m using the Hess screen with tangential correction of the grid for that distance. The separation and angulation of the streaks represent the subjective angle of the horizontal, vertical, and torsional deviations in a particular diagnostic position. To examine the responses of the fellow eye the flashlights are switched. The graphs from the Hess-Lancaster red and green streak test provide an excellent demonstration of and a useful diagnostic tool for incomitant strabismus. The smaller graph depicts the primary deviation, and the larger graph depicts the secondary deviation (see Fig. 7-2, A and B).

This test is quickly done and easily interpreted, and it is applicable to a cooperative child as well. However, the test does not distinguish between heterophoria and heterotropia. As with any subjective diplopia test, it indicates the dissociated position of the eyes without giving any information about the state of fusion. Because the distance between the two streaks represents the subjective angle, only when the patient has *normal retinal correspondence* (NRC) does this angle coincide with the objective angle (Fig. 3-5). By apply-

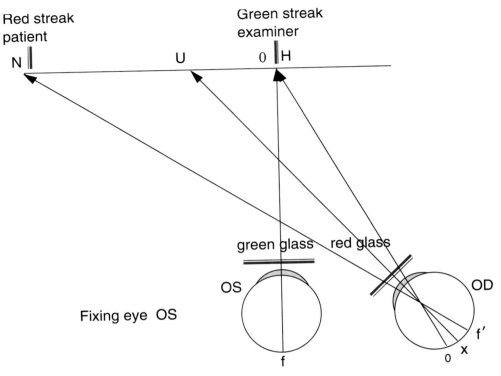

FIGURE 3-5. Principle of the Hess-Lancaster red-green streak test that presents to each eye a different target to be superimposed. The separation between the two streaks will correspond to the objective angle only in the presence of normal retinal correspondence (NRC). It will be less with abnormal retinal correspondence (ARC) unharmonious. There will be no separation between the streaks with ARC harmonious. Red streak superimposed at N-normal correspondence, objective angle; at U-Abnormal correspondence, unharmonious; and at H-abnormal correspondence, harmonious.

ing Prentice's rule (see Chapter 1), the angles of deviation can be calculated. Usually this is not done with the original manual method because of the time that such quantification requires.

Computer-assisted test. Hatch and Gonzalez[17] and Gonzalez, Hatch, and Sanchez[14] have described a computer-assisted Lancaster test (CALT). Their device uses a colored, liquid crystal display panel projecting the red and green streaks onto a digitized calibrated screen at 2.4 m in the nine diagnostic positions of gaze. The patient controls a three-position computer mouse to orient and superimpose the red and green stripes as seen through standard color-saturated red-green goggles, reversing them for measuring the fellow eye. The computer software, written and compiled in Quick Basic, collects, calculates, and stores the data in files that can be reproduced in either graphic or tabular form.

With this program the horizontal and vertical deviation in prism diopters as well as torsional deviations in degrees in each of the nine diagnostic positions are quickly recorded quantitatively and graphically. The instrument, which can be operated by one individual, has the advantage that the quantitative data obtained can be accurately and reproducibly transferred for computation and storage on computer disks. The coefficient of variation of 100 repeated measurements in subjects with prismatically induced horizontal deviations was 1.2% and with vertical ones, 4.5%. For patients with incomitant deviations and NRC, this test may replace the more demanding prism and cover test, which to be as accurate should be done at the same distance in all the diagnostic positions with each eye fixating in turn.

AC/A ratio measurements. Since Fry[11] described the concept of a ratio between accommodation and convergence, and Fry and Haines[12] in 1940 introduced the abbreviation AC/A for the ratio of accommodative convergence to accommodation, its evaluation has become an established part of the strabismus work-up. The two methods most often used clinically rely on the response to the prism and alternate cover test with the patient wearing the full refractive correction.

The *heterophoria method* calculates the ratio by dividing the difference between the near and distance measurements by the distance at near fixation in diopters (3 diopters) or by multiplying it by the distance at near fixation (1/3 m). The number obtained is added to the interpupillary distance in centimeters. The formula is as follows:

$$AC/A = PD + \frac{\Delta n - \Delta d}{3}$$

where PD represents the interpupillary distance in centimeters, Δn is the deviation at near, and Δd is the deviation at distance.

The *gradient method* uses +3 or −3 lenses added to the optical correction at a fixed distance. The examiner must ensure that the patient makes an effort to see the target clearly. Sometimes this is better achieved with less plus lenses or with minus lenses. The use of both plus and minus lenses allows a comparison. The formula of the gradient method is as follows:

$$AC/A = \frac{\Delta 1 - \Delta 0}{D}$$

where $\Delta 1$ is the deviation with the added lenses, $\Delta 0$ is the deviation without the added lenses, and D is the power of the lenses in diopters.

The gradient method more accurately estimates the AC/A ratio because it does not allow proximal convergence to affect its computation.

Effect of PAT on AC/A ratio measurements. In my experience, the AC/A ratio measurement may be different if taken after a PAT. Using the heterophoria method, a change in the figures can be expected if the prismatic correction has provoked a buildup of the deviation at near, for example, in an intermittent exotropia. However, a substantial difference may be found as well using the gradient method. *In patients for whom a low or high AC/A ratio will have a bearing on the choice of the muscle to be operated on and the amount of surgery to be performed, I recommend repeating the AC/A ratio measurements after the PAT with the prismatic correction still in place* (Case Presentation 3-1).

- **Case Presentation 3-1**

  Patient I.M.: This 9-year-old boy had a left exotropia, suppression left eye at distance, and alternate suppression at near.

  Exam:
    First measurement:
    ACPT: LXT 35Δ, LXT' 35Δ.
    Three months (after part-time patching OD):
    ACPT: LXT 30Δ, LXT' 30Δ, −3.00 X' 8Δ.
    AC/A: 7/1.
    Six months (under prismotherapy):
    ACPT: XT 35Δ, X(T)' 30Δ, −3.00 X' 6Δ.
    AC/A: 8/1.
    Nine months (presurgical measurements):

  PAT (stable one hour):
    XT 38Δ,   XT' 38Δ, −3.00   XT' 32Δ.
    AC/A: 2/1.
                           +3.00   XT' 48Δ.
    AC/A: 3/1.

  Surgery under microscope:
    Recession LLR 7 mm.
    Resection LMR 6 mm.

  Follow-up:
    One week: X 4Δ,   ⊕'.
    One month: flick X,   flick X'.
    Six months: flick X,   ⊕'.

  Comments:
    Because a normal to low AC/A was found with the PAT, the surgical plan was changed to a more appropriate recession-resection procedure rather than the bilateral lateral rectus recession that had been indicated initially by the high AC/A found before PAT was done.
    N.B.: The data presented are limited to the effect of PAT on AC/A. The results of prismotherapy are presented and discussed in Chapter 5.

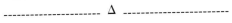

  Simultaneous prism and cover test. The simultaneous prism and cover test is used to evaluate a deviation under normal viewing conditions and not to reveal the total angle. This is well demonstrated in small-angle strabismus where the angle will "build up" with a prolonged alternate cover and prism test. For the simultaneous cover and prism test, the preferred eye is covered while a prism is applied simultaneously over the deviating eye. No movement will occur in the eye behind the prism when the angle has been neutralized *quickly.*

  Prism and corneal reflections. In the young child with inadequate coopera-tion and in the patient with little vision in one eye or *with eccentric fixation,* the deviation can be estimated by matching corneal light reflections by means of a prism (Krimsky Test).[24;25]

A hand-held light is positioned for the patient to fixate at near. To avoid parallax errors, the examiner directly observes the deviating eye. The power of the prism is changed until the corneal reflection in the deviating eye appears to have the same position as that of the fixating eye.

To facilitate observation, the prism is placed in front of the fixating eye. In large and in combined deviations, a prism may also have to be placed in front of the deviating eye. Fresnel trial set prisms facilitate the test in large-angle deviations (Fig. 3-6, A and B).

If the Krimsky test is performed on an orthophoric patient who is looking into the distance while a light is held at 33 cm, an 18Δ base-in prism will be needed to match the corneal reflections. Therefore if this method is used in a patient with strabismus, the measurement will have to be corrected by adding 18Δ to an esodeviation or by subtracting 18Δ from an exodeviation. A pa-

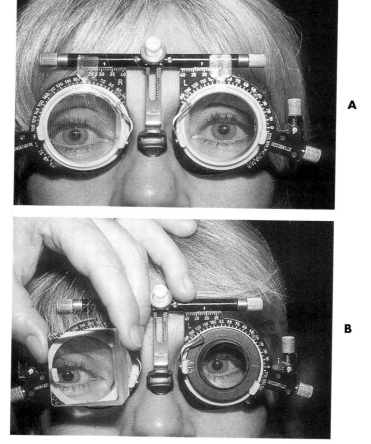

FIGURE 3-6. Fresnel trial set prisms facilitate the observation of the corneal light reflections. **A,** Centered with Fresnel trial prisms. **B,** With conventional prisms of same power.

tient looking into the distance with centered corneal reflections from a light held at 33 cm has, in fact, an esotropia of 18Δ. Krimsky described a perimeter type of instrument (the cardinal anglometer) to calculate the angle of strabismus in the diagnostic positions of gaze using his method.[23] If such an instrument were automated and computerized it could be very useful.

Although the Krimsky test is easy to perform in the infant, especially using the light behind a red glass, it may be difficult in the young child who usually opposes having any object placed in front of the eyes. In these cases the angle of deviation may be estimated by the displacement of the corneal light reflection alone (Hirschberg test). While holding a light at 33 cm in front of the fixating eye, the examiner estimates the displacement of the corneal reflex in the deviating eye. Each millimeter corresponds to about 8° (16Δ). The *angle kappa,* formed by the intersection of the visual axis and the central pupillary line, has to be taken into account. The angle kappa is habitually nasal (positive), and its average size of 5° (10Δ) is according to the calculation of Donders.[7] Smaller mean values have been reported.[9] There is also no agreement regarding the influence of eye length and refraction on the size of the angle.[8;13] With a positive angle kappa, the displacement looks less than it is in esotropia and more than it is in exotropia; it is the reverse with a negative (temporal) angle kappa.

Bielschowsky head-tilt test. In the presence of a vertical deviation, this test is done routinely. It consists of neutralizing the vertical deviation with prisms while the head of the patient is tilted toward the right or the left shoulder.[5]

If there is an associated horizontal deviation, it is also corrected; *the axis of the horizontal prism is maintained parallel to the floor of the orbit. The vertical prism is held with its axis perpendicular to the floor of the orbit* (Fig. 3-7).

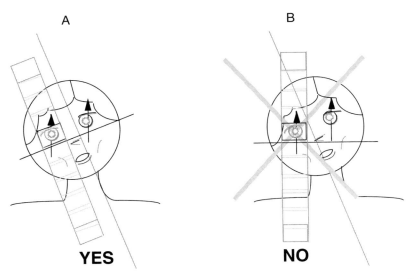

A   B

YES   NO

FIGURE 3-7. A, Correct position of prisms during the head-tilt test. Axis of vertical prism held perpendicular to orbital floor. (Axis of horizontal prism would be held parallel to orbital floor.) **B,** Incorrect position.

When the patient is sitting upright, a tilt of the head provokes a postural reflex movement to maintain the upright position of the environment. The eye on the side of the tilt intorts, that is, the upper pole of the cornea rotates nasally. The fellow eye extorts, that is, the upper pole of the cornea rotates temporally (Fig. 3-8). The response disappears when the patient is in the prone position.

Because the superior oblique and the superior rectus are intortors, and the inferior oblique and inferior rectus are extortors, the response to the head tilt can be analyzed for the four vertical muscles of each eye (Fig. 3-8). For example, with a paralysis of the right superior oblique, if the head is tilted toward the right shoulder, the right superior oblique will not contract, and the right superior rectus will then have to induce the intorsion of the right eye. Because it is a relatively weak intortor, its contraction will tend to be exaggerated to compensate the loss of intorting action by the paralyzed superior oblique. In addition, the vertical action of the superior rectus and the superior oblique no longer oppose each other. Because the simultaneous contraction of the superior rectus, an elevator, and the superior oblique, a depressor, normally induce only an intorsion of the eye, if the superior oblique is paralyzed, the elevating action of the superior rectus is no longer counteracted, and the paretic eye moves noticeably upward.

As Bielschowsky originally observed, in the presence of a superior oblique palsy, the vertical deviation increases when the head is tilted to the side of the palsy. *The response has a diagnostic value only when forced ductions are negative.* A fibrotic superior rectus on the same side can give a similar response.[34] A subjective test (e.g., diplopia evaluation with a Maddox rod) helps the interpretation in borderline cases.

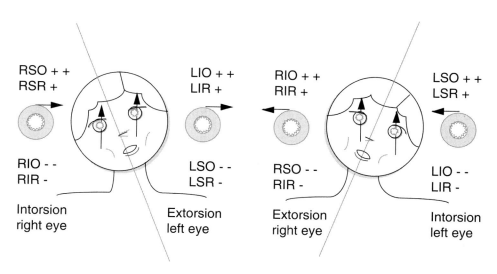

FIGURE 3-8. Postural reflex. Schematic of head tilt for analysis of four vertical muscles of each eye. Intorters: superior oblique and superior rectus. Extorters: inferior oblique and inferior rectus.

Maddox rod test. Maddox[27] designed a disc formed of parallel red- or clear-glass cylinders that project a series of punctiform images, the juxtaposition of which produces a linear image perpendicular to the axes of the cylinders. When a point light source is viewed through the disc it gives the appearance of a luminous streak. The disc is made to be hand held, inserted in a trial frame, or incorporated in a phoropter.

The Maddox rod test is especially useful in the evaluation of diplopia. The image it produces is more easily appreciated by the patient than that of a red glass. It is often used to confirm incomitance found with the prism and cover test and facilitates the head-tilt test. It can also verify the presence or absence of a torsional component in a deviation, which the red glass cannot. However, the red glass is less dissociative than the Maddox rod and is preferable to it for fusion testing.

To perform the test, the patient is asked to fixate on a light. The Maddox rod is placed in front of the nonpreferred eye, or if that eye is suppressing, in front of the preferred eye. When a horizontal deviation is evaluated, the disc is placed with the direction of the cylinders oriented horizontally to produce a vertical streak. For evaluation of a vertical deviation, the direction of the cylinders should be vertical to achieve a horizontal streak. If the streak induced does not intersect the fixation target light, the type of diplopia is determined and appropriately oriented prisms are placed over the Maddox rod until the streak and the light coincide. In combined deviations the larger deviation is corrected first. The primary compensating prism is kept in place while the associated smaller vertical or horizontal deviation is corrected. The influence of one or the other can be quickly reestimated. When neutralizing a horizontal deviation at near *in oblique directions of gaze, the Maddox rod must be held perpendicular to the fixation target so that it does not induce a false vertical phoria.*[30] In a patient with NRC, a false phoria is suspected when the Maddox rod test does not correspond to the cover test. The measurements are repeated with the Maddox rod and the prism bar placed over the fellow eye. If the streak remains strictly vertical with the cylinders at 180° and horizontal with the cylinders at 90°, with either eye fixating and in the diagnostic position of gaze, the patient has no cyclodeviation.

The power found corresponds to the subjective angle of a strabismus. The test does not differentiate between a phoria and a tropia. A patient with a heterophoria, when corrected, may report that the streak jumps from one side of the light to the other without bisecting the light. This patient has a suppression scotoma, the extent of which can be determined (see later section, Binocular Suppression Scotoma).

Cyclodeviation (torsional deviation) measurements. *Prisms cannot measure torsion.* The location of the fovea in relation to the optic disc can be used to document an intorsion or extorsion of the eye. However, it is more practical for the examiner to rely on the patient's subjective responses to the position of a linear image displaced along the sagittal axis. A light viewed through a Maddox rod provides a linear image that can be rotated to that position. After prism correction of associated vertical or horizontal deviations, if the streak is seen tilted, the patient or the examiner rotates the disc to redress the streak to a vertical or horizontal position. For example, if the patient sees the image intorted, the patient has an excyclodeviation, and the rod cylinders will be rotated outward. The examiner can read the degree of extorsion directly from the trial frame.

Some authors prefer to use two Maddox rods: red over one eye, and white over the fellow eye. The vertical deviation is undercorrected or overcorrected with appropriate prisms, leaving a small separation between the two horizontal streaks for easier identification of their relative positions. One streak is maintained in the horizontal position while the patient is asked to rotate the other streak until they are parallel. Both streaks may be seen tilted if the patient has a cyclodeviation in both eyes. In this instance the patient is asked to redress both streaks to a horizontal, parallel position.

The Hess-Lancaster red and green test also uses a streak to determine the presence of a cyclodeviation in different directions of gaze. In a traumatic double fourth nerve palsy with negligible associated deviations, it can reveal graphically an overlooked cyclodeviation responsible for the symptoms of a patient who may have been thought to be malingering.

The major amblyoscope, or a similar instrument, facilitates the evaluation of the binocular state with both cyclodeviation and associated deviations corrected, especially if central disruption of fusion is suspected.[29] However, in the majority of cases the single Maddox rod is both adequate and easily available for the diagnosis and measurement of a cyclodeviation. An automated, computerized Maddox-rod type of instrument should perform the test with greater facility and accuracy and produce a printout of all three deviations with either eye fixating in turn at a predetermined distance in the diagnostic positions of gaze.

No evaluation of vertical deviation with prisms can be considered complete unless the presence or absence of a cyclodeviation has been documented. In my experience, in many old deviations, cyclofusion compensates for a cyclodeviation of up to 15° that may be disregarded by the patient. However, in recent paralysis, a patient may complain of diplopia even with a 5° tilt. With time and the correction of associated deviations with prisms, cyclofusion intervenes, and the patient becomes asymptomatic although the cyclodeviation is still measurable. For this reason, prismotherapy should always be considered in the management of cyclodeviations (see Chapter 7).

Vergence measurement with prisms

Amplitude of motor fusion. The term motor fusion refers to the ability to align the eyes in such a manner that sensory fusion can be maintained.[28(p12)] The range of vertical and horizontal motor fusion can be measured with prisms by noting the power of the prism that provokes diplopia. Cyclovergences cannot be studied with prisms. *For this method to be valuable, either there must be no suppression, or it must be recognized should it intervene.* The patient with suppression will not see double and therefore will not make a compensatory movement of the eyes. This can be checked with a quick cover-uncover test that will not break fusion, if it is indeed present.

The optotype chosen for fixation should be slightly larger than the one re-solved by the eye with less vision and should still be seen through high-power prisms. The patient is asked to try to maintain clear single vision while prisms of increasing power are passed in front of the fixating eye. For a smooth tran-sition of powers, a prism bar or a rotary prism should be used.

To measure *horizontal amplitude of motor fusion*, the examiner starts with base-in prisms to evaluate the amplitude of divergence (negative fusional ver-gence), first at 6 m (D) and then at 33 cm (D′). Afterward, base-out prisms are used to evaluate the convergence (positive fusional vergence), again, first at distance (C) and then at near (C′). If base-out prisms are used first, the patient may have trouble relaxing convergence movements and may falsely show a poor divergence. The power of prism that provokes diplopia without the ability to recover (break point) is recorded. This power is then decreased progressively, and the maximum power with which the patient is again able to fuse is recorded. For example, if a 12Δ base-in prism provokes diplopia at 6 m, and a 10Δ base-in prism allows recovery, divergence amplitude will be recorded as D 12/10 (the break point as the numerator and the recovery point as the denominator).

It is important to record the recovery point because it indicates whether the patient regains fusion easily or with difficulty once it is broken. The pa-tient usually reports first that the optotype becomes blurred (blur point) be-fore becoming double. This occurs because accommodation is exercised to promote a greater degree of convergence, which happens shortly before the diplopia point. If there is too much discrepancy between the blur point and the break point, the former is considered to represent the "practical limit."

Divergence at near or distance is about one-third that of convergence when measured with prisms. The normal values for an orthophoric patient are ap-proximately:

$$D = 8/6\Delta \qquad D' = 12/10\Delta$$

$$C = 20\text{-}25/18\Delta \qquad C' = 35/30\Delta$$

Larger measurements are obtained with the amblyoscope, and divergence is found to be about one-fourth that of convergence. The sum of convergence and divergence values represents the range of fusion at distance and at near.

Although by convention the values are recorded as if the patient is or-thophoric, in their interpretation the heterophoria of the patient has to be con-sidered. For example, if a patient has an esophoria of 20Δ and a break point with a 2Δ base-in prism, this patient in fact has a divergence amplitude of 22Δ. This is important in deciding the treatment indicated for the symptom-atic patient (see Chapter 5). It is also important to ensure that the patient has been cooperative and to repeat the test before the final diagnosis.

Vertical amplitude of motor fusion is measured in a similar fashion by us-ing base-down and base-up prisms. To obtain the amplitude of fusion in sur-sumvergence or supravergence (right eye higher), the prism is placed base-down in front of the right eye or base-up in front of the left eye. For the am-plitude of fusion in deorsumvergence or infravergence (right eye lower), the prism is placed base-up in front of the right eye or base-down in front of the left eye. The two measurements will differ if the patient has a vertical deviation. For example, if the patient has a right hypophoria, in deorsumvergence (right eye lower), the amplitude will be found greater than in sursumvergence by about the amount of the phoria. Normally, the vertical amplitude of fusion is 3Δ to 6Δ. However, patients with long-standing vertical deviations or deviations with a mechanical factor (as in thyroid ophthalmopathy) may develop abnormally large vertical amplitude of fusion.[15]

Determination of Type of Retinal Correspondence

Determining the type of retinal correspondence is important to the management of motility disturbances. If a patient has abnormal retinal correspondence (ARC), a PAT is not done before surgery (see Chapter 4). Prisms are used only to determine the safer angle that can be operated without provoking diplopia. If a child with esotropia has ARC, prismotherapy is not undertaken (see Chapter 5), but decompensated strabismus with ARC may benefit greatly from prisms (see Chapter 7). Some patients spontaneously normalize their sensory state after surgery. How can such a good outcome be retrospectively verified if no meaningful testing has been done before surgery? More than one test is necessary to establish the type and depth of an anomaly. It may be the only way to determine the early onset of strabismus in an adult patient who complains of diplopia, does not provide old photographs, and denies a history of strabismus.

The afterimage test is not used in combination with prisms because they cannot influence the patient's answer. It is the test most removed from the casual seeing condition. When abnormal it reveals a deeply rooted anomaly.

Afterimage test. The afterimage test is readily used, even in children as young as 4 years of age, and is widely used to diagnose ARC. Afterimages are induced either by requesting the patient to look at the flash used to induce the afterimage or by the examiner positioning the afterimage at a specific point on the retina as with the euthyscope.[19(p486)] It is the former method that is described. A small, manageable afterimage test unit can be easily and economically made up by the practitioner.[19(p394)] It consists of a 110-volt strobe camera flash with its lens covered lengthwise by two strips of tape, leaving a fine slit between them. A small sector at the center of the slit is left opaque and colored to facilitate fixation.

Before beginning the test the examiner ensures that the fixation is foveal in each eye. The flash unit is held with the slit oriented horizontally. The patient is asked to fixate on the center target with the nondeviated eye, while the deviated eye is covered and the unit is activated. The fixating eye is then covered in turn, the slit is turned vertically and held in front of the deviated eye, and the strobe is fired again. The vertical position routinely is done last in front of the nonfixating eye because an amblyopic eye can better appreciate the vertical afterimage, and it will last longer. With the eyes closed, the afterimages are seen as positive, bright streaks on a dark background or as negative, dark grey or colored against a light background.

Afterimages are best viewed under intermittent room lighting (alternately switching the ambient light off and on). When the afterimage begins to fade it can be reactivated by gentle pressure on the globe through the lid.

If the foveas are corresponding points, the patient reports seeing a cross with a gap at its center, irrespective of the type and size of the deviation (Fig. 3-9, C). The gap at the center of each streak corresponds to the fovea excluded by the opaque target during the firing of the strobe. If the afterimages are seen not to match at their centers the patient has a deeply rooted ARC (Figs. 3-10, C, and 3-11, C) more so if the test is abnormal with positive and negative afterimages. Normal correspondence with afterimages does not preclude eliciting ARC with other tests.

With prisms alone. Prisms allow a point-by-point exploration of the retina of the deviated eye. If the patient fuses when corrected with prisms, and the cover test shows that the eyes are straight or in a heterophoric position, the patient has NRC (Fig. 3-9). If the cover test shows a residual tropia, the patient has ARC (unharmonious), and the subjective and objective angles will be different (Fig. 3-10). The difference between the two measurements is the angle of anomaly. For example, a patient fuses with a 20Δ base-out prism. A cover test over the right eye shows that the patient still has a residual left esotropia of 10Δ. For this patient the objective angle is 30Δ, the subjective angle is 20Δ, and the angle of anomaly is 10Δ (difference between the subjective and objective angle). If the patient fuses without any prismatic correction, while remaining esotropic, the patient has harmonious ARC (Fig. 3-11). The angle of anomaly

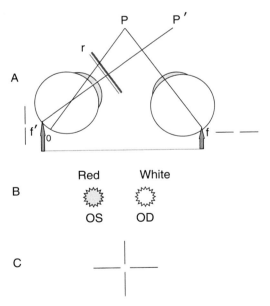

FIGURE 3-9. Normal retinal correspondence (NRC). **A,** f and f′ correspond. Subjective angle = objective angle. **B,** With diplopia tests, if there is no suppression or if it has been overcome with a red filter (r), the patient reports uncrossed diplopia corrected with a prism power equal to the angle of deviation. **C,** With the afterimage test stimulating both foveas the patient sees a cross.

equals the objective angle. This is the type of ARC most often found, especially with small-angle esotropia. In a patient with amblyopia and eccentric fixation, there is little indication for this test, as ARC is the rule.[19(p163)]

Red glass and prisms. In patients with larger angle deviations or suppression, the type of diplopia and retinal correspondence are more easily determined with a red glass placed in front of the preferred eye. This method, described by von Graefe in 1855,[16(p14)] is the oldest still in use for diagnosing ARC. The clinical information from the test often duplicates that of the major amblyoscope (Synoptophore or Troposcope). Prisms and a red glass have the advantage of being more readily available.

First, the type of diplopia is determined without any prismatic correction. Therefore the relationship examined is that between the fixating fovea and the point in the deviating eye where the image of the object fixated by the nondeviating eye falls. (Conventionally, this point is called point zero.) Second, the relationships between the fovea of the nondeviating eye and the area between point zero and the fovea of the deviating eye are studied by correcting the diplopia with a prism bar passed in front of the red glass. Patients with ARC may report paradoxical diplopia (against the rule), that is, crossed diplopia when still esotropic (see Fig. 3-10), uncrossed diplopia when exotropic, a higher diplopia image with a hyperdeviation, or a lower diplopia image with a hypodeviation. When uncrossed diplopia is reported, the examiner

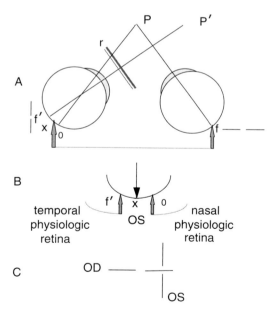

FIGURE 3-10. Abnormal retinal correspondence (ARC), unharmonious. **A,** f and x correspond. The subjective angle is greater than zero but less than the objective angle. **B,** With diplopia tests the patient reports uncrossed diplopia up to a point when crossed diplopia is reported, although the esodeviation is undercorrected with prisms. **C,** With the afterimage test the patient reports crossed diplopia. The angle of anomaly is less than with harmonious ARC.

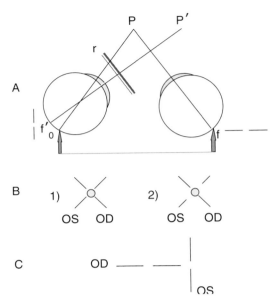

FIGURE 3-11. Abnormal retinal correspondence (ARC), harmonious. **A,** f and 0 correspond. The subjective angle is zero. The angle of anomaly equals the objective angle. **B,** Typically with Bagolini lenses with no correction the patient reports fusion without (B1) or with (B2) a central scotoma OS. Patient also reports fusion with the red glass. **C,** With the afterimage test the patient reports crossed diplopia. The left afterimage in the physiologic temporal retina is localized to the right.

must ensure that the patient is not in fact alternating. Paradoxical diplopia is a sure indication of ARC. However, it is rare that a patient with ARC would spontaneously complain of paradoxical diplopia. It may happen after a change, generally postoperative, in the objective angle. Although it has been reported, with rare exceptions, to be a fleeting phenomenon limited to the immediate postoperative period,[28(p258)] it can be disturbing and can be predicted or avoided by a PAT done before surgery (see Chapter 4).

Patients with ARC (unharmonious) initially report diplopia "with the rule" using the red-glass test (Fig. 3-10). For example, an esotropic patient reports uncrossed diplopia with the red glass. Base-out prisms of increasing power are passed in front of the red glass up to the point where diplopia is overcome. At that point the cover test reveals that the patient is still esotropic and has ARC. At the objective angle, this patient, unless suppressing, will report crossed diplopia as the image falls now on the fovea of the deviating eye, which has adopted a temporal localization. Surgical overcorrection without modification of ARC will also allow diplopia "with the rule" (Fig. 3-12). A careful examination with the prism bar and the red glass will differentiate this diplopia from that of NRC. The patient may also report binocular triplopia (monocular diplopia of the deviated eye) if normal and abnormal correspondence are present. The patient perceives one image from the nondeviated eye and two from the deviated eye, one image being localized by point x and the other

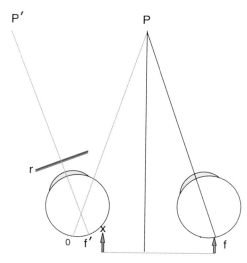

FIGURE 3-12. Surgical correction without modification of ARC allows diplopia "with the rule." As described in Fig. 3-8, f and x are still corresponding. With the red glass (r) the patient will report crossed diplopia as long as point 0 remains temporal to x. NOTE: point x is not found outside zone f '0 unless the strabismus has been treated habitually by surgery.

by the fovea. In the past this phenomenon has been induced purposely as a treatment of ARC.[19(p565)] but is no longer favored. It is rarely spontaneous, and attention should not be directed toward it.

Worth 4-dot test. The Worth 4-dot test uses a pair of glasses fitted with color-saturated red and green lenses through which the patient views four dots—one red seen through the red glass, one white seen by each eye as red or green or superimposed, and two green seen through the green glass—the colored dots are seen projected by a flashlight at 33 cm and at 6 m from a light box or projected on a screen. The outer circumference of the four lights of the test sustains an angle of 6° at 33 cm. At distance the angle is 1.25° or less, depending on the instrument used. The answers that the test provides under these dissociating conditions are analyzed as follows:

1. Two or three dots indicate suppression of one eye. The glasses are then reversed. Depending on the type of ametropia and the antisuppression effect of the red glass, the patient may report fusion or diplopia.
2. Four dots indicate bifoveal or peripheral (outside suppression scotoma) fusion. The patient with microtropia and harmonious ARC typically reports fusion at near and suppression at distance.
3. Five dots: If the examiner is certain that the patient is not quickly switching fixation, the five-dot response (diplopia "with the rule") may be that of a dissociated heterophoria, an intermittent tropia for which prisms can be used to achieve fusion. It can also indicate an unharmonious ARC (see Fig. 3-10). *Paradoxical diplopia is always associated with ARC.*

Although this test can more quickly pinpoint the diagnosis than the Bagolini lenses, and especially in the younger age group, it is also more dissociating.

Bagolini striated lenses. Bagolini striated lenses are available in a two-pair boxed set with striations of different widths and density. These lenses are constructed from clear micro-Maddox type of cylinders that are thin enough to allow an objective view of the patient's eyes. The streaks that the patient sees projected from the fine striations, while looking at a light, provide a diplopia test as close as possible to casual seeing conditions.[2] Similar striated lenses fixed in a frame to prevent inadvertent rotation are commercially available but are not as versatile as the boxed set.

To perform the test the examiner selects from the set the pair of lenses with the finer striations. If the patient has trouble distinguishing the lines with both eyes, the pair of lenses with the broader striations is substituted. One lens is placed over the left eye so that the luminous streak is seen at an angle of 45°. A second lens is placed over the right eye with its striations perpendicular to the those of the first so that the streak is seen at an angle of 135°. The lenses are placed to project the lines obliquely so that their representation on a diagram is not confused with the horizontal-vertical charting for the afterimage test. However, it should be impossible to confuse the two tests because their principles are different. The afterimage test is done only with bilateral foveal fixation, with the flash stimulating the fixation point of each eye. The Bagolini lenses apply the principle of the diplopia tests, with one luminous streak being seen by the fovea of one eye and the other being seen by the point in the deviating eye where the image of the nondeviating eye falls (see Figs. 3-9, 3-10, and 3-11).

When the lines are seen in diplopia, the subjective angle is determined with prisms up to the point at which the lines cross to become an X. If the cover test shows that a tropia remains, it is neutralized with prisms to determine the objective angle of the strabismus (Fig. 3-13).

The patient may suppress one line completely or only a part of it around the fixation light, demonstrating a typical central suppression scotoma with peripheral fusion (see Fig. 3-11, B2). The extent of the scotoma is studied by rotating the line.

In most cases of early onset small-angle strabismus ($\leq 15\Delta$), the two striations are fused at near without any prismatic correction while a manifest deviation is demonstrated with the cover test. These patients have a *harmonious ARC*. As the subjective angle is 0, the angle of anomaly is equal to the objective angle (see Fig. 3-11).

Polaroid vectograph tests. *Near stereoacuity* is commonly measured clinically with the Titmus stereotest that is composed of Polaroid vectograph plates imprinted in such a way that certain characters or objects are polarized at 90° to others, resulting in an image disparity. When viewed through Polaroid spectacles at 40 cm, each image presents as a separate target to each eye and assumes a different position in space. The apparent difference between the im-

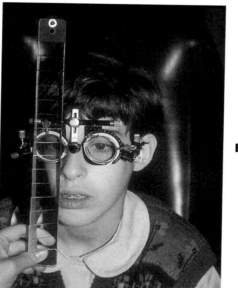

FIGURE 3-13. Bagolini lenses for fusion testing, OD at 45°, OS at 135°. **A,** In a child's trial half frame. **B,** In combination with a prism bar. **C,** In combination with Fresnel trial set prisms correcting the angle.

ages is measured in seconds of arc. The highest disparity a patient can detect with the test at 40 cm is 40 seconds of arc at the ninth circle target, and the lowest, measured with the housefly target at the same distance, is 3000 seconds. In between are three rows of animals, imaging disparately at 400, 200, and 100 seconds of arc, and nine circles with thresholds from 800 to 40 seconds of arc. To avoid monocular clues, the test booklet is opened only after the patient has been equipped with the polaroid spectacles. The test is conducted at 40 cm under good light, avoiding reflections with the booklet held squarely because any tilt will compromise the test by altering the axis of polarization.

I consider a patient who continues to show 40 seconds of arc with repeated testing, and who sees the test objects receding with the booklet held upside down and loses stereopsis with either eye covered, to be bifoveal. In microtropia the test may show as much as 100 seconds of arc and up to 60 seconds of arc with the monofixation syndrome. Although the test may allow monocular clues, an erroneous response should be readily differentiated from true stereopsis when the test is properly administered. The Lang test, designed so the patient does not have to wear glasses, is especially useful in the young child. It presents image disparities ranging from 600 to 1200 seconds of arc.[28(pp266-267)]

Distance stereoacuity is commonly measured clinically with the projected vectograph test (American Optical Vectographic Project-O-Chart slide) imaged on a nondepolarizing aluminized screen at 6 m. The patient wears polarized glasses and identifies which target, if any, in each of the five rows of test circles appears to be three dimensional. In the top row the disparate target subtends 240 seconds of arc with the disparities decreasing downward through four successive rows, each with a randomly placed target circle subtending 180, 120, 60 seconds of arc and in the bottom row 30 seconds of arc. Stereopsis is tested *after fusion* has been demonstrated with the Project-O-Chart fusion slide, which projects two polarized lines that point in different directions according to the viewing eye: to the left and down for the left eye and to the right and up for the right eye. The patient who fuses sees the lines converge as a cross. This test can be used in combination with prisms and is, with similar tests that apply the same principle, *the most simple way to evaluate fusion and stereopsis at distance before, during, and after treatment.* The same slide projects optotypes—that can be seen only with the right eye, only with the left eye, or with both eyes—that are used to detect a suppression area or malingering.

Major amblyoscope. The major amblyoscope, the Synoptophore, the Synoptiscope, and the Troposcope are similar instruments that derive from the Worth amblyoscope. Some models are designed to produce afterimages and Haidinger's brushes. This *biocular* instrument attempts to create the condition of distance vision. Through its respective ocular, each eye sees an illuminated slide that can be moved independently in vertical, horizontal, and torsional directions. The slides are designed to test simultaneous perception, fusion, and stereopsis. The description of the instruments and the interpretation of each test, as well as applicable therapeutic methods, have been covered extensively in Hugonnier's book.[19] In North America, because of the poor therapeutic results achieved, the major amblyoscope or similar instruments are rarely used by ophthalmologists or orthoptists to treat ocular motility disturbances. Vergence exercises have been an exception. However, the motivated and well-instructed patient can perform them as effectively at home or at work without any instrument or with prisms at a significant saving of time and cost (see Chapter 5). The major amblyoscope has lost much of its popularity as an instrument for routine motility examinations. It is used in the child who will cooperate to be evaluated with an amblyoscope after refusing other tests. It is useful also to detect fusion and stereopsis with the torsional, vertical, and horizontal deviations corrected. Notwithstanding, its evolution, in a computer-generated format, is currently being promoted as a 21st Century therapeutic as well as a diagnostic tool for binocular anomalies.

Duality of retinal correspondence with various tests. It is possible for the patient to give different answers regarding retinal correspondence in different directions of gaze, by changing eye fixation or with different tests.[3;16(pp15-18);18(pp186,196-199)] *The less dissociative the test or the closer it is to everyday vision, the more easily ARC is elicited.* In his classical paper, Bagolini found that in 98 cases of esotropia, only 35 had ARC when tested with a flash-induced streak afterimage test, whereas 64 demonstrated it with the synoptophore (12 harmonious and 52 unharmonious), and 84 cases had harmonious ARC with striated lenses.[19(p199)] The incidence of this anomaly, as revealed with red glass and prism testing, falls between that of afterimages and striated lenses testing. ARC with afterimages (negative and positive) is deeply rooted.

It is also possible that, using the same test, the answers differ according to the control of the deviation. This duality of retinal correspondence is better illustrated in intermittent exotropia. Some patients with manifest exotropia and ARC will demonstrate NRC with the same test when the exotropia is controlled.

Binocular suppression scotoma. With amblyopia a central scotoma (absolute scotoma) may be present monocularly but when either eye is covered, suppression, as manifested in the binocular condition, disappears (relative scotoma). Its extent varies greatly according to the test used, the duration of the test, attention span, and surrounding conditions. Except for patients with alternating exotropia and suppression of the entire retina of one or the other eye in turn, suppression involves only a portion of the retina (scotoma). Thus a prism of a given power will allow the light rays to fall outside the scotoma, and if the eye remains motionless, the patient will report diplopia. By changing the power and direction of the prism, the size of the suppression scotoma can be measured. Prisms provide a simple method to objectively evaluate suppression clinically.

To approximate conditions of everyday vision, the patient is seated in a well-lit room to look at a fixation light 6 m away. If the patient with strabismus does not see double, the patient is either suppressing or inattentive. In esotropia, base-out prisms are passed in front of the deviating eye. If, with a given power, the patient sees double, the examiner has reached a nonsuppressed area of the retina. Using vertical prisms, the vertical extent of the scotoma can be determined as well.

The dimensions of the scotoma vary also according to the conditions of the test. The scotoma found when moving from the suppressed area toward the diplopia area will be smaller than that found when moving from the diplopia area toward the suppressed area.

A colored glass placed in front of the deviating eye will cause the scotoma to diminish greatly or disappear entirely.

Depth of suppression. The depth of suppression can be estimated with red filters of different densities. The greater the suppression, the darker the filter necessary to provoke diplopia. Bagolini[1;4] has designed a ladder of red filters with a scale of 1 to 17 (Fig. 3-14). Number 16 still allows the perception of a light. Number 17 is almost a total occluder. The red glasses commonly found in trial sets correspond to numbers 9 to 12 of the Bagolini ladder. Although the filter works primarily by reducing illumination to the fixating eye, it also works in another way. With the lighter red color of the scale, up to number 8, diplopia may be provoked by applying the filter over the deviating eye as well as by applying it over the fixating eye. Also, Bagolini recommends putting the filter in front of the deviating eye to avoid a change in the pattern of fixation.[4] If diplopia is elicited before surgery at the objective angle with prisms alone, indicating no suppression at that angle or with a light-colored glass indicating a mild suppression, it will be present after surgery if orthotropia or overcorrection is achieved. On the other hand, if diplopia is provoked only with the red glass or with a filter of higher density, postoperative diplopia is not probable. Nevertheless, this possibility cannot be discarded completely in the presence of deep amblyopia.[6] The secure zone for surgical correction can be estimated by determining the size of the suppression scotoma.

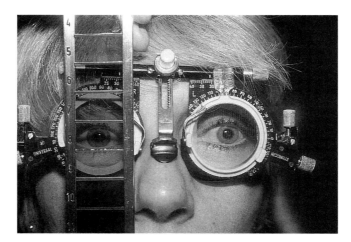

FIGURE 3-14. Bagolini bar of red filters used in combination with Fresnel trial set prisms to determine the depth of suppression and the secure zone for surgical correction.

The 4Δ base-out test. The 4Δ base-out test, described by Irvine[20] and popularized by Jampolsky,[21;22] is used to detect a small suppression scotoma under binocular conditions. The introduction of a base-out prism in front of one eye normally induces a version movement of both eyes in the direction of the apex of the prism. This is followed by a corrective movement of the eye without the prism (see Fig. 3-3).

The test is carried out with the patient fixating on a light or a single optotype at distance. The examiner introduces a 4Δ base-out prism in front of one eye and observes the induced response.

1. If both eyes make a version movement followed by a recovery movement of the eye without the prism, there is no suppression scotoma.
2. If both eyes make a version movement but the eye without the prism does not make a recovery movement afterward, this eye has a foveal suppression scotoma.
3. If there is no version movement of either eye, the eye behind the prism has a suppression scotoma.

Atypical responses to the test have been observed[32;33] and the examiner should use caution when making a diagnosis solely on the basis of a 4Δ base-out test.[10] Usually a recovery movement in both eyes indicates bifoveal fusion, and the test is termed negative for a microtropia.

REFERENCES

1. Bagolini B: Presentazione di una sbarra di filtri a densità scalare assorbenti i raggi luminosi, *Boll Ocul* 36:638, 1957.
2. Bagolini B: Tecnica per l'esame della visione binoculare senza introduzione di elementi dissocianti (test del vetro striato), *Boll Ocul* 37:195-209, 1958.
3. Bagolini B: Diagnostic errors in the evaluation of retinal correspondence by various tests in squints. In Arruga A, editor, *International strabismus symposium, Giessen, 1966*, Basel, 1968, S Karger, pp 163-174.
4. Bagolini B: Usefulness of filters of progressive density as a diagnostic tool for some strabismic problems. In Reineke R, editor: *Strabismus. II. Proceedings of the Fourth International Strabismological Association, Asilomar, California, 1982*, New York, 1984, Grune & Stratton, pp 561-566.
5. Bielschowsky A: Lectures on motor anomalies of the eyes. II. Paralysis of individual eye muscles, *Arch Ophthalmol* 13:33-59, 1935.
6. Campos EC, Cavallini GM, Codeluppi R: Weakness of anti-diplopic mechanisms favours post-operative diplopia in patients with esotropia and amblyopia, In Murube J, editor: Madrid, 1988, Acta XVII Concilii Europaeae Strabologicae Associatis, pp 65-72.
7. Donders FC: *On the anomalies of accommodation and refraction of the eye*, London, 1864, The New Syndenham Society. (Translated by Moore WD.)
8. Effert R, Gruppe S: The amount of angle alpha in a normal population, In Murube J, editor:, Madrid, 1988, Acta XVII Concilii Europaeae Strabologicae Associatis, pp 17-22.
9. Franceschetti AT, Burian HM: "L'angle kappa," *Bull Mém Soc Fr Ophtal* 84:209, 1971.
10. Frantz KA, Cotter SA, Wick B: Reevaluation of the four prism diopter base-out test, *Optom Vis Sci* 69:777-786, 1992.
11. Fry GA: Further experiments on the accommodation-convergence relationship, *Am J Optom* 16:125, 1939.

12. Fry GA, Haines HF: Taits' analysis of the accommodation-convergence relationship, *Am J Optom* 17:393, 1940.
13. Giorianni FG, Siracusano G, Cusmano R: The angle kappa in ametropia, *New Trends in Ophthalmol* 3:27, 1988.
14. Gonzalez C, Hatch JF, Sanchez RM: Computer-assisted Lancaster test. In Program of The Am Assoc of Ped Ophthalmol and Strab, 19th Annual Meeting April 17-21, 1993, Session III, Abstracts for Posters, Westin Mission Hills Resort, Palm Springs, Calif, p 20, 1993.
15. Greaves BP, Mein, J, Gibb JC: Long-term follow-up of patients presenting with dysthyroid eye disease, In Moore S, Mein J, Stockbridge W, editors: *Orthoptics: past, present, future. Transactions of the third International Orthoptic Congress, Boston, 1975*, New York, 1976, Stratton Intercontinental Medical Book Corp, pp 223-240.
16. Halldén U: Fusional phenomena in anomalous correspondence, *Acta Opthalmol Kbh Suppl* 37:1-93, 1952.
17. Hatch JF, Gonzalez C: Computer quantitation of the classic Lancaster test, Scientific Poster 244, *Am Acad Opthalmol*, 147, Nov 1992, (scientific program).
18. Hugonnier R: *Strabismes, Hétérophories, paralysies oculo-motrices— les désiquilibres oculo-moteurs en clinique*, Paris, 1959, Masson et Cie.
19. Hugonnier R, Clayette-Hugonnier S: *Strabismus, heterophoria, ocular motor paralysis, clinical ocular muscle imbalance*, St Louis, 1969, Mosby, (Translated and edited by S. Véronneau-Troutman).
20. Irvine SR: A simple test for binocular fixation: clinical application useful in the appraisal of ocular dominance, amblyopia ex anopsia, minimal strabismus and malingering, *Am J Opthalmol* 27:740, 1944.
21. Jampolsky A: The prism test for strabismus screening, *J Pediatr Ophthalmol* 1:30, 1964.
22. Jampolsky A: Symposium on strabismus. In Allen J, editor: *Transactions of the New Orleans Academy of Ophthalmology*, St Louis, 1971, Mosby, pp 65,66.
23. Krimsky E: The cardinal anglometer, *Arch Ophthalmol* 26:670-674, Oct 1941.
24. Krimsky E: The fixational corneal light reflexes as an aid in binocular investigation, *Arch Ophthalmol* 30:505, Oct 1943.
25. Krimsky E: The effect of a prism on the corneal light reflex, *Arch Ophthalmol* 39:351, March 1948.
26. Lancaster WB: Detecting, measuring, plotting and interpreting ocular deviations, *Arch Ophthalmol* 22:867-880, Nov 1939.
27. Maddox EE: A new test for heterophoria, *Ophthalmic Rev* 9: 129-133, 1890.
28. von Noorden GK: *Binocular vision and ocular motility: theory and management of strabismus*, ed 4, St Louis, 1990, Mosby.
29. Pratt-Johnson JA, Tillson G: Acquired central disruption of fusional amplitude, *Ophthalmology*, 86:2140-2142, 1979.
30. Putnam OA, Quereau JV: Precisional errors in measurement of squint and phoria, *Arch Ophthalmol* 34:7-15, 1945.
31. Risley SD: A new rotary prism. *Trans Am Ophthalmol Soc* 5:412-413, 1889.
32. Romano PE, von Noorden GK: Atypical responses to the four-diopter prism test, *Amer J Opthalmol* 67:935, 1969.
33. Roundtable discussion. In Gregersen E, editor: *Transactions of the European Strabismological Association*, Kopenhagen, 1984, Jencodan Tryk, p 215.
34. Véronneau-Troutman S: A four-step test for diagnosis of pseudo superior oblique palsy, *Graefes Arch Clin Exp Ophthalmol* 226:317-322, 1988.

CHAPTER 4

Prism Adaptation Test

History

For as long as ophthalmic prisms have been used for the diagnosis and treatment of ocular motor disturbances, before prescribing them, a trial to determine their efficacy has been advocated. Prisms have been used diagnostically as well to determine the sensory condition and the angle of strabismus before surgery.[39;21] However, the prism adaptation test (PAT) described by Jampolsky in 1971,[24] giving credit to Fletcher Woodward an orthoptic technician (1965), had a more defined goal. The PAT test was *specifically* advocated to determine preoperatively the reaction of the esotropic patient to a slightly overcorrecting base-out prism. Not only was it limited to esotropic patients, but also it was intertwined with the use of prismotherapy. The following three possibilities were considered:

1. If the patient, to secure bifoveal fusion, made a vergence movement to compensate a slightly overcorrecting prism, the angle found was the angle operated on.
2. If the patient made no vergence movement and diplopia resulted from the overcorrecting prism, the prism was worn for 1 hour. If the condition remained unchanged at the end of the hour, a prism of the same power was prescribed. Following more prolonged wear, if reaction (1) took place, treatment was carried out accordingly.
3. The patient reestablished an esodeviation. This "overconverging adaptation phenomenon" was thought to indicate that the patient would do the same after surgery and return to the original angle of esotropia.

To prevent a return to the original angle, even in the child with very early onset esotropia, therapy to break through suppression and anomalous retinal correspondence (ARC) before surgery was strongly advocated. Many techniques to treat ARC, including prisms, were in vogue at that time.* Jampolsky advocated a technique that used overcorrecting prisms, taking advantage of the new Fresnel prism membranes, after the methods of Bagolini[7] or Cüppers,[1;2] and that involved a very long period of treatment.

---

*References 2-5, 7, 11-13, 22,35.

Efficacy of Prism Adaptation Test

Little prospective data was available regarding the usefulness of the prism adaptation test as it was originally described until *the multicenter prospective randomized study published in 1990.*[36;37] It was designed from the original description of PAT with two exceptions. It did not include children with onset of esotropia before 6 months of age. Patients with ARC were not treated or were handled differently from patients with NRC and no fusion. About 10% of the esotropia patients screened met the stringent eligibility requirements of this study: age of 3 years or older, onset of esotropia after age 6 months, no previous treatment except for amblyopia and spectacle correction within 0.50 diopter of the cycloplegic refraction, residual concomitant esotropia of 12Δ to 40Δ, vertical component not more than 3Δ, distant and near deviation differential not more than 10Δ (with or without bifocals), no dissociated vertical deviation (DVD), no nystagmus, no fusion at the objective angle (fusion at the subjective angle was accepted: ARC), and finally, the patient had to be a surgical candidate.

Technique used in the multicenter PAT study. Patients in the multicenter PAT study were fitted with Fresnel prism membranes to offset the angle of esotropia found at distance with the alternate cover and prism test (ACPT). The power was divided equally between the two eyes or favoring slightly the nonfixating eye. At weekly intervals for 4 weeks the prismatic correction was adjusted for an increase in the esotropic angle not to exceed 60Δ.

At the end of 4 weeks, of the 199 prism-adapted study patients, 131 were found to have a residual esotropia at distance and near between 0Δ to 8Δ (simultaneous prism cover test), to fuse the Worth 4-dot test at 33 cm, or to report diplopia with this test and to identify at least two circles and two animals on the Titmus stereoacuity test. These patients were called responders, and they were divided into two groups. Sixty-seven patients underwent surgery for the original angle of deviation as determined with the ACPT. Sixty-four patients underwent surgery for the PAT angle as determined with the *simultaneous prism cover test.* Surgery was done according to a strict protocol.

Results of the multicenter PAT study. Six months after surgery, 89% of the responders operated on for the *adapted angle* had an angle of deviation ranging from 0Δ to 8Δ. Of this group, 69% demonstrated fusion. Of those operated on for the *original angle,* 79% had a similar correction of which 61% demonstrated fusion. Statistically, with 6 months of follow-up, there was a trend toward better motor and sensory results when the response to PAT was incorporated into the surgical treatment of acquired nonaccommodative esotropia. In particular, when the subgroup of responders with a built-up angle of more than 10Δ was analyzed separately, the motor results 6 months after surgery were significantly improved. A successful alignment was achieved in 89% (24 of 27 patients) with prism-adapted surgery versus 70% (21 of 30 patients) with surgery at entry.

Comments. Probably more responders would have been found to have a built-up angle over 10Δ if their deviation had been determined with the ACPT (method used for all other groups) instead of the simultaneous prism cover test. Unfortunately, patients with fusion and NRC were discarded from this prospective study on the presumption that they had no potential to benefit from prism adaptation. In my experience, as well as that of others,[19;44] these patients can manifest a substantial angle buildup with prisms and benefit from "augmented surgery."

Because of the strict protocol of this rigorously conducted study, some questions remained unanswered. It is intriguing that so many patients developed some binocular link after wearing a prismatic correction for only *1 month or less*, and this occurred even after the age of 8. It is also surprising, knowing the side effects of high-power Fresnel membranes (see Chapter 2) that the older age group tolerated powers above 15Δ over both eyes for such a long time. Since patients with ARC as well as with NRC were accepted for PAT, it would be interesting to know what percentage of the responders had ARC and the effects of PAT on that group.[45]

It would be instructive to know the sensory and the motor outcomes of the responders with small-angle esotropia included in the study. Jampolsky, who originally thought that the overconverging adaptation phenomenon was a sign that the patient would do the same after surgery and return toward the original angle of esotropia, is now of the opinion that the multicenter PAT study has shown the following:

> "A person may converge to a prism to a lesser residual angle, and then with a stronger overcorrecting prism verge some more to a still smaller residual angle. If this repeated prism trial results in a *small angle* esotropia (5Δ to 8Δ) it is then quite a stable condition. One could then operate and obtain a good result, i.e. a *small angle* esotropia. It is the same with a fusion response(BFF). Both conditions are very stable. It may not matter how much surgery is done, as long as enough is done. It may not have to be for the total angle found with PAT, (as the study showed)—but the Multicenter Study Group demonstrates again—importantly—that once you get a small angle esotropia it's stable and you can drag it around 'almost' the same as BFF (Bifoveal fusion), with surgery, and in my (Jampolsky's) experience, with slow, decremental changes in the amount of prism."[25]

In Bagolini's opinion, "the suggestion of the Prism Adaptation Study Group to operate the angle found after prism compensation is wrong. They do operate not only the basic deviation but also an innervational part which may change with time."[10]

Ohtsuki study. Ohtsuki et. al.[34] published a prospective study of preoperative prism adaptation (5 to 7 days) in 77 patients with nonaccommodative, acquired esotropia after 6 months of age and with angles of deviation ranging from 18Δ to 50Δ. These patients did not have dissociated vertical deviations, manifest or latent nystagmus, vertical deviation larger that 3Δ, A and V patterns, visual acuity less than 0.50, high AC/A ratios, and previous surgery. Of the 77 patients, 63 were found to be responders, that is, while wearing prisms their angle at distance was within 10Δ with the ACPT, and all fused with the Bagolini lenses. Those whose angle had built up more than 60Δ or who demonstrated a suppression response were classified as nonresponders. The responders were randomly assigned to two groups, one set of 32 patients underwent surgery according to the original angle before PAT, and the second set, 31 patients, underwent surgery for the prism-adapted angle. At 1 year follow-up, although there was a trend in favor of the responders who had surgery for the adapted angle, even with an increase of $\geq$ 8Δ the difference in the motor and sensory results achieved was not significant.

Similarities and differences between the two studies. Like the Prism Adaptation Study Research Group, patients in the Ohtsuki study were corrected for their distance deviation. Kutschke et al. have since advocated correction of the near deviation.[27] Contrary to the multicenter PAT study protocol, in the Ohtsuki study all angles of deviation were evaluated with the ACPT. The Bagolini lenses were used instead of the Worth 4-dot test to ascertain fusion during the adaptation. However, either test may allow a positive response with microtropia in the presence of harmonious ARC, which in turn allows gross stereopsis. The time allowed for the adaptation was shorter, 5 to 7 days instead of 4 weeks. The angle of deviation before PAT was 18Δ to 50Δ compared to 12Δ to 40Δ in the Prism Adaptation Study Research Group, and had a smaller patient sample. The follow-up to achieve the final result was longer, 1 year instead of 6 months. In the responder groups the motor success rate was similar in the two studies. The multicenter PAT study had 89% with augmented surgery versus 79% with surgery at entry. The Ohtsuki study had 84% with augmented surgery versus 78% with surgery at entry. In the multicenter PAT study the observed benefits of prism adaptation were concentrated mainly in patients who built-up to larger angles and who underwent surgery for their adapted angles. In the Ohtsuki study such a conclusion could not be made. Nonresponders had worse motor results (50% = 0Δ to 10Δ ET or XT) than the nonresponders of the multicenter study (73%). They had a history of an earlier onset and a longer duration of their strabismus. This is an important factor, as emphasized by Lang[28] in his discussion of the results of the Prism Adaptation Study Research Group.

Comments. It is interesting that in the *nonresponder* groups in both studies, fusion that had *not* been detected before surgery was found after surgery. In the Prism Adaptation Study Research Group of the 67 nonresponders, 73% were aligned (0Δ to 8Δ) of whom 34% demonstrated fusion. In the Ohtsuki study of 14 nonresponders, 7 had good motor results, 3 with fusion. Fusion has been observed to develop with time once parallelism has been achieved. If fusion was immediate after surgery, it may be explained by the use of high-power Fresnel prism membranes for PAT[43] (see Chapter 2). When corrected with these membranes, nonresponders with large-angle deviations may have had fusion although they did not acknowledge it. *Hard plastic Fresnel trial lenses have to be used for the evaluation of the motor and sensory state for angles of 40Δ and greater.* Fresnel trial prisms, available in powers up to 40Δ without the optical disadvantages of membranes above 20Δ, allow PAT to be performed more efficiently and accurately for large-angle deviations. On the other hand, their appearance limits them to office use.

Influence of Retinal Correspondence on PAT

Decrease of esotropia with PAT. In 1970 at the First Congress of the International Strabismological Association (ISA), Aust and Welge-Lussen[5] gave their results of prism compensation (adaptation) in esotropia. Bérard,[14] Pigassou-Albouy,[35] and Wybar[47] also participated in the symposium dedicated to the use of prisms in the preoperative and postoperative treatment of strabismus. Interestingly, during the same symposium Jampolsky, Flom, and Thorson[23] presented for the first time their paper "The Membrane Fresnel Prism, a New Therapeutic Device," and gave the basis for the PAT.

Aust and Welge-Lussen were also attempting to verify the theory of Adelstein and Cüppers, who had explained a relapse of esotropia with ARC by "a return to the acquired abnormal cooperation in the earlier objective angle."[1;5] Using 5 to 9 days of prism compensation before surgery, they attempted to determine this tendency preoperatively.

Their retrospective study included 88 esotropic children of ages 5 to 8, maintained as inpatients for close observation, with concomitant deviations, bilateral equal vision, no or negligible vertical element, an angle up to 56Δ, and with an optical correction for hypermetropia within 0.50 diopter as determined by atropine refraction. The age of onset was not given, but the retinal correspondence was carefully studied. ARC was found in 46 patients (52%). The prism compensation averaged between 5 and 9 days with twice daily control.

The angle increased in 63 patients (71.5%), and in 22 of these (25%) it increased by 12Δ or more. In 14 patients (16%) the angle remained unchanged. In 11 patients(12.5%) the angle decreased (7 with ARC and 4 with NRC). There was no difference in the angle buildup with PAT in patients with ARC and NRC. Postoperatively with a follow-up of 1 year or more, no relationship could be found between the sensory state and the postoperative angle variations.

In 14 patients (30.4%) the correspondence became spontaneously normal after surgery, in 5 of them the angle had increased by 12Δ or more with PAT, in 6 patients the increase was less than 12Δ, and in 2 the angle had decreased. Postoperatively, 5 patients were divergent, 5 were slightly convergent, and 4 were parallel before correspondence became normal. Aust and Welge-Lussen concluded that postoperative divergence did not lead to a change of correspondence more often than did a slight convergence.

After a 1-year follow-up the group of patients operated for their angle buildup did not have more postoperative divergence than those whose angle did not increase. This was also found with a longer follow-up.[46] However, of those who decreased their angle with PAT, two thirds had a relapse of their esotropia of from 8Δ to 24Δ. Although only 11 patients (12.5%) reduced their angle, the percentage of relapse is high and would lead to the recommendation to operate for the total deviation.

At the present time, Aust uses prism compensation 3 to 7 days before surgery, not only in esotropia but also in exotropia and vertical strabismus. He operates on the built-up angle except when diplopia is present at that angle. He also does not operate for a very small angle that has shown continuous buildup with no stabilization.[6]

Etiology of angle buildup with PAT

Patients with abnormal retinal correspondence. Patients with abnormal retinal correspondence (ARC), especially when associated with small-angle strabismus, may increase their angle of deviation when corrected with prisms. Since its first description by Travers,[42] this response has intrigued many strabismologists.* It cannot be explained by a type of diplopia-phobia mechanism because many patients with such a buildup do not see double. Another explanation is that the buildup is induced to maintain the old angle of anomaly. For Bagolini these fusional movements are always anomalous and associated with ARC. They express a variation of the retinal motor value. He explained the observation that very small-angle strabismus and microtropia return to their original preoperative angle, whereas larger angles do not because of the ARC point-to-area relationship, which is small and deeply rooted in microtropia and large and less well defined in large-angle esotropia.[8;9] In patients with ARC, Halldén observed four decades ago that there is not always an angle buildup with prism compensation. He attributed the minimal to no change in the angle to a rapid change in the angle of anomaly.[20]

Patients with normal retinal correspondence. These patients not only have normal retinal correspondence (NRC) but also may have normal stereoacuity (Titmus test: 40 sec of arc at 40 cm).[19;33;44] The buildup provoked by prism neutralization may be marked and found in esodeviation as well as in exodeviation. In esodeviation a buildup caused by a convergence spasm will be accompanied by myosis and pseudomyopia. With this etiology eliminated, one may speculate that prisms allow these patients to reveal the full angle, which had been masked by divergence efforts in esotropia and convergence efforts in exotropia. This explanation does not suffice as demonstrated by the fact that otherwise similar angles would be reached with diagnostic occlusion. When the buildup achieved by the two methods was compared, although some authors found it to be equal,[40] others found it greater with PAT.[19;29] Some authors have theorized that prisms, which allow the patients to look with both eyes simultaneously, may change the pattern of the afferent impulses reaching the oculogyric centers, which alter efferent impulses, causing the increase of the angle.[19]

It is possible that the buildup in both ARC and NRC has a common etiol-

*References 1, 5, 8, 9, 16-20, 30.

ogy unrelated to the type of retinal correspondence. On the other hand, clinically the buildup is interpreted differently according to the type of retinal correspondence. In patients with NRC the total angle induced by PAT is interpreted as the true deviation on which the patient is habitually operated. In patients with ARC the buildup seems to be a reaction of "protest," artificially created; this becomes especially apparent in *small-angle strabismus*. In patients with ARC, my practice is not to operate on an angle buildup induced either by a prolonged ACPT or PAT.

Evolution of the PAT Indications and Technique

Esotropia with ARC. I do not perform the PAT in esotropic patients with ARC for two reasons:

1. Except for small-angle esotropia at distance, esotropic patients with ARC usually do not return to their original angle after surgery.[32;26] PAT may lead to inappropriate surgical indications in this group. A barely noticeable deviation may appear to require a surgical correction because of the buildup. Likewise a persistently repeated ACPT can provoke a similar misleading response.

2. Finally, prismotherapy alone or combined with other orthoptic methods only rarely cures the anomaly (see Chapter 5). Although such an outcome is not frequent, spontaneous "near normalization" after restored parallelism by surgery is the best result that can be achieved.

In the adult with known ARC the test is done only to predict postoperative diplopia. However, in a child, where insufficient cooperation does not allow the confirmation of ARC with other tests and an early onset cannot be ruled out, repeated angle buildups without orthophoria being achieved at any time are interpreted as an ARC response.[41] Prismotherapy, as well as other therapeutic attempts to establish bifoveal fusion, is not undertaken. Except for the treatment of amblyopia and the compensation of an accommodative component with spectacles, the child, as is the adult patient with esotropia and ARC, is managed according to the size of the angle and the cosmetic blemish ascertained *before PAT.*

Esotropia with NRC. If the patient makes a vergence movement to compensate a slightly overcorrecting prism so as to secure bifoveal fusion, the test is continued, and the original angle is not operated on as recommended in the original description of PAT. On the contrary, the adaptation is allowed to continue, and the prismatic correction is increased to compensate any angle buildup. The new angle must remain stable for at least half an hour before the adaptation is considered complete. Patients with NRC and perfect stereopsis at near and distance may markedly increase their angle of deviation. A borderline phoria-tropia may become a clear surgical indication. It is impossible to predict who will respond with a significant angle buildup. When operated on for the angle found with PAT, these patients will have better results.[19] For this reason PAT should be done routinely in patients with NRC.

All strabismus with normal fusion potential. If PAT becomes synonymous with a presurgical prism trial test to reveal the full angle of strabismus in patients with normal fusion potential, it follows that it should be used with exo-deviations as well.[38;44] The presence of a vertical deviation should not be a contraindication to the test. Many years ago it was recognized that very often a vertical deviation will disappear once the horizontal deviation is corrected with prisms. This response is taken as an indication not to operate for the vertical component at the primary surgery.[26;31;44] PAT will also provide useful information about incomitant deviations, a V or A pattern, and nystagmus (see Chapters 5, 6, and 7).

One objection to the systematic use of a prism trial test to reveal the full angle of strabismus in this group of patients is the absence of prospective studies showing better surgical results. It is also based on the false assumption that patients with normal fusion potential do well despite the amount of surgery done. Prospective studies are the best method to determine the significance of promising anecdotal or retrospectively analyzed techniques. However, to my knowledge there are no published prospective studies showing that surgical results are not as good when based on measurements estimated from corneal reflections than when based on more exact measurements made with the prism and cover test, or with the fixation target at a constant distance at near and in the far gaze in the diagnostic positions. Nevertheless, the latter two techniques continue to be recommended because it seems logical to think that more accurate measurements applied preoperatively may enhance the surgical outcome and postoperatively facilitate an objective evaluation of the results.

On the other hand, the use of PAT must be differentiated from that of prismotherapy, for which a different approach is required. For example, in a patient with NRC under prismotherapy, the minimum power that controls the deviation is prescribed. Prisms are placed in such a way as to alter the eye preference and overcome suppression (stronger prism over the dominant eye). However, when prismotherapy is completed, if surgery is indicated, it is performed for the full angle of deviation found with PAT. This time the prisms are applied so as not to change the eye preference (equal prisms or the weaker one over the dominant eye).

Author's Technique

PAT as presurgical test in a patient with fusion potential and NRC. The specific application of PAT and its surgical implications are detailed in each section on ocular motility disturbances. In general the technique that I use is as follows:

1. In the office the patient wears prisms over the indicated optical correction (including bifocals if needed) to correct the horizontal and vertical angle of strabismus as determined by the ACPT. Clear round prisms are used for angles below 20Δ, and *prisms from the Fresnel trial set* are used for larger angles. As a rule, oblique prisms are not used to facilitate es-

timation of the different powers for presurgical angle evaluation. Fresnel membranes are not used.

2. Initially, if the angle differs at distance and near, the larger deviation is neutralized. If constant diplopia persists at near or distance, the prismatic correction is reduced to allow single vision and increased *gradually* afterwards. It is my experience that in exotropia, with rare exceptions, the initially smaller near angle builds to approximate the larger deviation at distance. However, in esotropia, except for masked phoriatropia, (see Chapter 5) frequently the near distance disparity remains.

3. Either the total prism power found is divided equally over the two eyes or the powers are selected and positioned in such a way as *not* to alter the eye preference. The powers of the prisms are changed according to variations in the angle found with the alternate cover test performed at 10- to 15-minute intervals. One-half hour after the angle has become *stable*, it is recorded at near and distance, and the *sensory testing is repeated with the prisms in place*, finalizing the PAT.

Duration of prism adaptation test. The duration of the PAT is very important to patient compliance. Few patients will accept wearing high-power prisms over both eyes for weeks at a time, although most patients will tolerate them in the office for 1 or 2 hours. Also, because of their definite optical advantages over membrane prisms (when the test is done in the office), regular prisms, wafer prisms, or Fresnel trial set prisms are better tolerated and should be used. In many studies, adaptation is considered completed by the end of 1 week. However, when combined with prismotherapy, the duration of the "test" can vary from weeks[9] to months.[15;35]

When used to determine the presurgical angle of a deviation(s), it has been my experience, as well as that of others,[19;41;44] that in most cases PAT can be completed in the office in 1 to 2 hours. The exception is in the adult patient with heterophoria-tropia or in an old decompensated deviation with diplopia when fully corrected by prisms, because of a tendency to return to an abnormal head posture or because of ARC. Initial undercorrection followed by gradually increasing prismatic correction, prescribed at follow-up visits and worn at home, will eventually elicit the full deviation that can be corrected surgically without inducing diplopia. However, in the patient with ARC, if the test provokes an angle buildup, it is interrupted. The patient is operated on for the subjective angle if it is cosmetically advantageous. The possibility of postoperative diplopia and the need for a prismatic correction have to be accepted, even if the diplophia was initially compensated with an adjustable suture.

Summary and Conclusions

The PAT as originally described involving treatment of ARC is rarely done. Its modification when rigorously applied prospectively by the Prism Adaptation Study Research Group[34;36;37] showed a trend overall and little statistical significance to support the test. On the other hand its abbreviation as PAT has been catching and has replaced other terminology applicable to prism tests as currently done. The prism adaptation test (PAT) has expanded to include all strabismus and nystagmus for which a response to a prism trial determines the surgical treatment.

REFERENCES

1. Adelstein FE, Cüppers, C: Probleme der operativen Scheilbehandlung, *Dtsch Ophthalmol Ges* 69:580, 1968.
2. Adelstein FE, Cüppers C: Le traitement de la correspondence rétienne anormale à l'aide des prismes, *Ann Oculist (Paris)* 203:445, 1970.
3. Arruga A: *Diagnóstico y tratamiento del estrabismo,* vol I:291, vol II:212, Bermejo, Madrid, 1961.
4. Arruga A: The use of space diagnostic methods and of prismotherapy in the treatment of sensory alterations of convergent squint. In *The First International Congress of Orthoptics, 1967,* London, 1968, Henry Kimpton, pp 62-76.
5. Aust W, Welge-Lussen L: Preoperative and post-operative changes in the angle of squint following long-term pre-operative prismatic compensation. In Fells P, editor: *The First Congress of the International Strabismological Association, Acapulco, Mexico, 1970,* London, 1971, Henry Kimpton, pp 217-226.
6. Aust W: Personal communication, June 1993.
7. Bagolini B: Postsurgical treatment of convergent strabismus with a critical evaluation of various tests. In Schlossman A, Priestly B, editors: *Int Ophthalmol Clin, 1966,* vol 6, no 3, Philadelphia, 1966, Little, Brown and Co, p 633.
8. Bagolini B: Sensorial-motorial anomalies in strabismus (anomalous movements), *Doc Ophthalmol* 41:23, 1976.
9. Bagolini B, Zanasi MR, Bolzani R: Surgical correction of convergent strabismus: its relationship to prism compensation, *Doc Ophthalmologica* 62:309-324, 1986.
10. Bagolini B: Personal communication, May 1993.
11. Baranowska-George T: L'hypercorrection prismatique dans les strabismes traités selon la métode de localisation, *Bull Soc Ophtalmol Fr,* 69:192, 1969.
12. Bérard PV, Carlotti S, Payan-Papera R: Les traitements orthoptiques dans l'espace des correspondances rétiennes anormales, *Bull Soc Ophtalmol Fr* 65:750, 1965.
13. Bérard PV: Prisms: their therapeutic use in strabismus. In Arruga A editor: *International Strabismus Symposium, University of Giessen, 1966,* Basel, 1968, S Karger, pp 339-334.
14. Bérard, PV: The use of prisms in the pre- and post-operative treatment of deviation in concomitant squint. In Fells P, editor: *The First Congress of the International Strabismological Association, Acapulco, Mexico, 1970,* London 1971, Henry Kimpton, pp. 227-234.
15. Bérard, PV, Reydy R, Berthon J: Permanent wearing of prisms and early delayed treatment of esotropia, In Moore S, Mein J, Stockbridge L, editors: *Orthoptics: past, present, future, transactions of the Third International Orthoptic Congress, Boston, 1975,* New York, 1976, Stratton Intercontinental Medical Book Corporation, pp 203-208.
16. Burian HM: Sensorial retinal relationship in concomitant strabismus, *Trans Am Ophthalmol Soc* 43:373, 1941.
17. Campos EC, Zanasi MR: Die anomalen fusions bewegungen der sensomotorische aspekt des anomalen binokularsehens, *Graefes Arch Clin Exp Ophthalmol* 205:101, 1978.
18. Carniglia PE, Cooper J: Vergence adaptation in esotropia, *Optom Vis Sci* 69:308-313, 1992.
19. Delisle P, Strasfeld M, Pelletier D: The prism adaptation test in preop

erative evaluation of esodeviations, *Can J Ophthalmol* 23:208-212, 1988.

20. Halldén U: Fusional phenomena in anomalous correspondence, *Acta Opthalmol Kbh Suppl* 37:1-93, 1952.

21. Hudelo A: La prescription des prismes en pratique courante, *Ann Ocul* 174:528-541, 1937.

22. Hugonnier-Clayette S: Traitement orthoptique de la correspondance rétienne anormale dans les ésotropies. Etude de 236 cas, *An Inst Barraquer* 12:417-430, 1974-75.

23. Jampolsky A, Flom MC, Thorson JC: Membrane Fresnel prisms: a new therapeutic device. In Fells P, editor: *The First Congress of the International Strabismological Association, Acapulco, Mexico, 1970,* London, 1971, Henry Kimpton, pp 183-193.

24. Jampolsky A: In Allen J, editor: *Symposium on Strabismus. Transactions of the New Orleans Academy of Ophthalmology,* St Louis, 1971, Mosby, pp 66-75, 354-360, 396, 397.

25. Jampolsky A: Personal communication, July 1993.

26. Knapp P: Use of membrane prisms, *Trans Am Acad Opthalmol* 79:OP718-OP721, 1975.

27. Kutschke PJ, Scott WE, Stewart SA: Prism adaptation for esotropia with a distance-near disparity, *J Pediatr Ophthalmol Strabismus* 29:12-15, 1992.

28. Lang J, Heinreich T: Prism adaptation in acquired esotropia, Correspondence, *Arch Ophthalmol* 110:751-752, 1992.

29. Langenthal-Schildwächter-von A, Kommerell G, Klein U, Simonsz HJ: Preoperative prism adaptation test in normosensoric strabismus, *Graefes Arch Clin Exp Ophthalmol* 227:206-208, 1989.

30. Maraini G, Pasino L: Variations in the angle of anomaly and fusional movements in cases of small-angle strabismus with harmonious retinal correspondence, *Brit J Ophthalmol* 48:439-443, 1964.

31. Moore S, Stockbridge L: An evaluation of the use of Fresnel Press-on prisms in childhood strabismus, *Am Orthop J* 24:62-66, 1975.

32. von Noorden GK, Muñoz M: Recurrent esotropia, *J Pediatr Ophthalmol Strabismus* 25:275-289, 1988.

33. von Noorden GK: *Binocular vision and ocular motility,* ed 4, St Louis, 1990, Mosby, pp 461-462.

34. Ohtsuki H, Hasebe S, Tadakoro Y, Kishimoto F, Watanabe S, Okano M: Preoperative prism correction in patients with acquired esotropia, *Graefes Arch Clin Exp Ophthalmol* 231:71-75, 1993.

35. Pigassou-Albouy R: The use of prisms in pre- and post-operative treatment, In Fells P, editor: *The First Congress of the International Strabismological Association, Acapulco, Mexico, 1970,* London, 1971, Henry Kimpton, pp 235-242

36. Prism Adaptation Study Research Group: Efficacy of prism adaptation in the surgical management of acquired esotropia, *Arch Ophthalmol,* 108:1248-1256, 1990.

37. Prism Adaptation Study Research Group: *Manual of procedures.* USPHS No 5UO1EYO5296, 1984.

38. Ron A, Merin S: The use of the pre-op prism adaptation test (PAT) in the surgery of exotropia, *Am Orthop J* 38:107-110 1988.

39. Sattler CH: Prismenbrillen zur Frühbehandlung des Konkomittierenden Schielens, *Klin Monatsbl Augenheilkd* 84:813-816, 1930.

40. Shippman S et al: Prisms in the preoperative diagnosis of intermittent exotropia, *Am Orthop J* 38:101-106, 1988.

41. Shippman S, Cimbol D, Weseley AC: The preoperative use of prisms in esotropic children, *Am Orthop J* 34:72-76, 1984.

42. Travers TB: *The comparison between the results obtained by various methods employed for the treatment of concomitant strabismus,* London, 1936, G Pulman & Sons.

43. Véronneau-Troutman S: Fresnel prisms and their effects on visual acuity and binocularity, *Trans Am Ophthal Soc* 76:610-653, 1978.

44. Véronneau-Troutman S: Surgical implications of the prism adaptation test, *Binocular Vis* 1:107-109, 1985.

45. Véronneau-Troutman S: Prism adaptation test (PAT) in the surgical management of acquired esotropia, Correspondence, *Arch Ophthalmol* 109:765-6, 1991.

46. Welge-Lussen L, Aust W: Changes in the angle of squint after prismatic correction: postoperative follow-up results. In *Disorders of ocular motility: neurophysiological and clinical aspects,* Symposium der Deutschen Ophthalmologischen Gessellschaft, 1977, pp 341-346.

47. Wybar K: The use of prisms in preoperative and postoperative treatment, In Fells P, editor: *The First Congress of the International Strabismological Association, Acapulco, Mexico, 1970,* London, 1971, Henry Kimpton, pp 243-249.

CHAPTER 5

Prismotherapy

Goals of Prismotherapy

Prisms, because of their optical properties, can be used to realign the visual axes *(motor effect)* and correct diplopia when present. They also can be used to facilitate the stimulation of corresponding retinal points to assist in the development of binocular vision in a patient with fusion potential *(sensory effect)*. As with many medical treatments, in a significant number of patients prismotherapy often corrects only the symptoms and rarely by itself achieves a cure. However, to neglect prismotherapy as a part of the total management of strabismus is as undesirable for the patient's welfare as to use it indiscriminately in all ocular motility disturbances.

There is a time to initiate prismotherapy and a time to discontinue it. When a child is unsuccessfully treated for an amblyopia and unable to attain binocular vision, I always inform the parents as well as the child that Babe Ruth, the legendary baseball player and "Home Run King" was deeply amblyopic in one eye. This revelation often helps them both to accept the less than anticipated sensory outcome of treatment and enables the family to treat the child as normal and not irreparably handicapped.[40]

Prisms in the Treatment of Amblyopia

Mild or residual amblyopia can be treated in conjunction with prismotherapy using the reduction in vision induced by the prism membrane as a partial occluder. The examiner must ensure that the membrane occludes the eye with the better visual acuity sufficiently to have fixation taken over by the amblyopic eye. Development of a binocular link will not only stimulate further improvement in visual acuity but also can prevent a relapse. Given the choice, the older child will often prefer a Fresnel membrane to total full-time or even part-time occlusion, rarely applied after age 12.

Treatment of amblyopia is indicated in small-angle strabismus in the child under age 7 wearing a full refractive correction. The age limit is extended to age 11 for the anisometropic myopic patient.[77] Patching should be stopped immediately if the angle begins to increase. If it is continued, the angle buildup may persist when discontinued.[72] Once the buildup is established, usually no prism can be found that stabilizes the angle to overcome the diplopia of which the young patient often bitterly complains. Decompensated microtropia in the adult responds better to prismotherapy (see Case Presentation, 7-10). For the child who does not return spontaneously to the original small-angle and sensory pattern, surgery is the treatment of choice (Case Presentation 5-1).

- **Case presentation 5-1**

Patient M.G.: This five-year-old child had a large-angle esotropia and intermittent diplopia. Past history revealed an *anisometropic amblyopia OS* detected at age 4½ with *no apparent deviation* (confirmed with photos), treated with glasses and patching for 4 to 5 months. Her visual acuity OS had improved to 20/30, but a large *right esotropia with diplopia* developed. Patching had been discontinued only recently.

First exam:
   Wears: OD plano = 20/50.
              OS − 4.25 = 20/25.
   ACPT: RET 40Δ,   RET′ 45Δ.
   Postcycloplegic refraction: OD +1.25 +1.25 × 130° = 20/30.
                                              OS − 3.00 −1.00 × 160° = 20/30.
   Glasses changed to take advantage of the bilateral equal vision for a trial of prism therapy or alternatively to allow a return to OD preference.
   After new correction for one week: ET 60Δ, ET′ 55Δ.
   Alternate suppression: distance and near.

Prismotherapy:
   30Δ base-out Fresnel membranes over each eye.

Follow-up:
Four months: no indication of fusion distance and near. Only slight
   decrease in deviation. Has adopted a left face turn. Prisms removed.
One year: visual acuity: OD (cc) = 20/20, OS (cc) = 20/30
   ACPT: (cc) ET 50Δ, ET′ 65Δ.
              (sc) ET 50Δ, ET′ 55Δ.
   Correction OS replaced by plano.
Six weeks later: LET 40Δ, LET′ 45Δ, as at her first visit but with OD
   preference.

Surgery under microscope:
   Recession LMR 5 mm.
   Resection LLR 8 mm.

Follow-up:
Three weeks: parallelism.
   Fusion at near (Worth 4-dot test).
   Suppression OS at distance.
Six weeks: near stereoacuity 200 sec of arc (Titmus).
Four months: gradual decrease in visual acuity OS over intervening period
   to 20/40 with finding of 100 sec of arc stereoacuity (Titmus).
One year: further decrease of visual acuity to 20/80 OS.
   Sensory state unchanged. 4Δ test positive for microtropia.

Four years: sensory and motor state unchanged. Fuses also the Bagolini striated lenses at near without suppression.

D 2Δ,   D′ 8Δ,   C 14Δ,   C′ 14Δ.

Comments: This case illustrates well how many sensory anomalies associated with microtropia are deeply rooted and important for the binocular link. Amblyopia was easily overcome and the eye preference changed, but an angle buildup resulted with the loss of binocularity. The amblyopia had to be allowed to relapse, and parallelism had to be restored to allow the original condition, microtropia with peripheral fusion and gross stereopsis, to become reestablished.

Contrary to our usual policy, bilateral Fresnel membranes of 30Δ were applied over both eyes for a prolonged period in this 5-year-old child because the parents were very anxious that another approach than surgery be attempted first.

Inverse prism in the treatment of amblyopia with centric and eccentric fixation. In 1965 Rubin described a method that used an inverse prism for treating small-angle strabismus with an amblyopia, not worse than 20/50, and with centric or eccentric fixation (unsteady central, parafoveal). The angle of esotropia or exotropia, primary or resulting from a slight undercorrection or overcorrection (also following early surgery), had to be 10Δ or less. For esotropia a 5Δ base-in prism was placed over the amblyopic eye or for exotropia, a 5Δ base-out prism. A calibrated occluder that reduced the visual acuity two lines below that of the amblyopic eye was applied over the sound eye. Of 118 cases of small-angle deviation, 59.3% reportedly became parallel, 43 of 54 cases with eccentric fixation became centric, with improved stereoacuity in 69 of 97 cases tested.[69]

Rubin recognized the objection to his technique, that the lateral rotation of an esotropic eye with a base-in prism is only temporary. The eye adducts to automatically refixate at the eccentric point. However, the sound eye has also rotated medially as the eye behind the prism has moved laterally. To resume foveal fixation, being only partially occluded, it abducts (see Fig. 3-3). Rubin postulated that the series of rocking movements made by the esotropic eye outward and inward by this mechanism would result in an increase in tonus of the lateral rectus that would eventually move the eye far enough outward to achieve foveal fixation.

Pigassou-Albouy, although using total rather than partial occlusion of the sound eye, claimed a similar mechanism of action and reported excellent results also duplicated by others.[49] She has published extensively on this form of treatment that she applied to any degree of amblyopia with eccentric fixation. In her latest publication she still recommended it.[56-61] In her original description the distance between the eccentric point and the fovea was estimated to determine the power of the inverse prism. Later she relied on the corneal reflex to ascertain the power, between 6Δ and 20Δ (Fresnel membrane or wafer), necessary to achieve a slight overcorrection.

Garcia de Oteyza and co-authors applied Pigassou's technique for 6 to 12 months to 70 esotropes, ages 18 months to 14 years. In 1988 they reported a 24% cure rate.[24] In the 31 patients with treatment after age 7, the visual acuity improved from 0.28 to 0.38, duplicating the poor results published in this age group with this as well as other methods.[8;70] Nevertheless, Garcia de Oteyza continues to apply the technique.[26]

As the different techniques advocated combined the use of an inverse prism over the amblyopic eye with penalization or partial or total occlusion of the sound eye, it would seem that any improvement in fixation and visual acuity cannot be attributed solely to the inverse prism. These authors have not reported a controlled study, which might have shown, as it did for pleoptics, that patching of the sound eye is in reality the only reason for improvement in visual acuity.[22;50;75]

Prisms for Heterophoria

Symptomatic phorias. The vast majority of the population has some degree of heterophoria. In most cases the deviation is kept latent without any conscious effort. As a rule, asthenopic symptoms must not be attributed to a "coexisting" heterophoria unless all other etiologies have been eliminated. If the symptomatic phoria began following a change of glasses, the cause could be an unplanned prismatic effect from poorly fitted frames or decentered lenses or the inappropriate inclusion of prisms. Additional causes, common to spectacles as well as to other means of refractive correction, have to be considered as well:

1. A change in eye dominance.
2. The correction of an old refractive error that had never been corrected.
3. A sudden increase in the accommodation-convergence requirement.
4. A change from spectacles to contact lenses or correction by refractive surgery or with an intraocular lens.

Because of the potential for the development of symptomatic phoria, *any patient planning to have refractive surgery for correction of an ametropia needs to be carefully evaluated preoperatively,* especially since refractive surgery correction is not readily reversed as with spectacles and contact lenses.

The lenses of single-vision spectacles are usually centered for the wearer's interpupillary distance at far gaze. Therefore at near the spectacle-corrected myope has a decentration-induced base-in prism effect that is not present when wearing contact lenses or after refractive surgery. In either case the demand for accommodation increases, and an exophoric myope can become symptomatic for close work. Conversely the hyperope, after refractive surgical correction or when wearing contact lenses, loses the base-out prism effect of a spectacle correction at near. The esophoric hyperope compensates this disadvantage by the reduced demand for accommodation.

To pinpoint the cause of a symptomatic phoria, the history, visual requirements, refraction, type of correction, near point of accommodation (NPA), and motility evaluation must all be determined with great care.

For the Hugonniers, a patient with symptoms from a decompensated heterophoria usually has a poor amplitude of fusion and a poor stereoacuity and is a candidate for orthoptics.[36] Other authors use prisms to correct the heterophoria. With time, as the sensory state improves, the patient becomes asymptomatic. Except in the older age group[27] symptomatic heterophoric patients with "true" poor fusion amplitudes are better treated first by orthoptics. If there is no substantial angle buildup with PAT and symptoms relapse because of the motor aspect of the condition, a prism prescription is considered.

Prism prescription. Because of the poor vertical amplitude of fusion, before a prism correction is prescribed, the classical teaching is to first fully correct any vertical phoria. This alone can render the patient asymptomatic and make a horizontal correction unnecessary. Also, some authors recommend to fully correct an esophoria.[9;44;45] Others undercorrect it by ⅓ to ½ to retain active fusional divergence and to prevent the patient from requesting prisms of increasing powers to remain asymptomatic. However, not all heterophoric patients react in this fashion. A PAT performed initially can target the patient who will respond with an angle buildup that, if large, contraindicates prismotherapy and indicates surgery.

Sheard's criterion is used by some authors to determine when to use a prismatic correction as an alternative to PAT. In this instance the prism power prescribed is calculated as ⅔ of the phoria minus ⅓ of the compensating duction. For an exophoria the compensating duction is the amount of base-out prism that will induce blur (or break if there is no blur) and conversely for an esophoria, the amount of compensating base-in prism. If the result is negative, no prism is needed.[46;84]

Phoria adaptation to prisms

Normal adaptation. Several authors have reported on the oculomotor system response of normal subjects to a small vertical or horizontal prism. Henson and North[31] have found that, following 3 to 5 minutes of either a vertically induced change of 2Δ or a horizontally induced change of 6Δ, the adaptation system operates to maintain the original level of fusional vergence and phoria position. They also reported[52] that patients with binocular problems often have a deficient adaptation system. Milder and Reinecke found a deficient adaptation in patients with cerebellar dysfunction.[48] Carter[16] and Ogle et. al.[53] have reported that patients with poor adaptation often obtain comfortable binocular vision with a prismatic correction.

Masked intermittent tropia adaptation. Patients who have a reasonable amplitude of fusion, normal stereoacuity at near and distance, normal near point of accommodation, and, only at first, small phorias with the alternate cover test and prism and with the Maddox rod, may build up their angle of deviation when wearing a prismatic correction. Such a large masked deviation can be responsible for their asthenopic symptoms with episodes of blurred vision or diplopia and reading problems, helped by closing one eye. A PAT performed initially, before prescribing a prism, would have revealed the full angle. An office trial will show that to relieve the symptoms of the majority of these patients requires a full prismatic correction, often too large to be incorporated. A few weeks' trial of a Fresnel membrane over the nondominant eye will confirm this. *Because these patients have normal retinal correspondence they should be operated on for the built-up angle.*

In masked tropias, to give a prismatic correction to compensate the original angle of deviation (before PAT) requires repeated increases in the prismatic correction. These patients are said to "eat up" the prisms. The misdiagnosis leading to inappropriate management of this group may be the cause for some authors to be against using prisms for heterophorias. *It is important not to confuse this group of bifoveal heterophoric patients with those with anomalies such as microtropia, ARC, peripheral fusion, and gross stereopsis that present a similar response to PAT.* Their management is completely different (Case Presentation 5-2).

- **Case presentation 5-2**

  Patient R.F.: This 36-year-old man had eye strain for close work and intermittent diplopia for the past 10 years with no history of trauma or illness. Previous treatment consisted of orthoptic exercises, bifocals for 2 years, and low-power incorporated prisms consecutively prescribed in the following powers 1.5, 4, 8, 4, 12, and 6Δ with only temporary relief of symptoms.

  First exam:
  Visual acuity (pc): OD −0.75 −0.50 × 180° = 20/20.
  OS −1.25 −0.50 × 180° = 20/20.
  NPA: normal, AC/A ratio: 4/1.
  ACPT (cc): E(T) 10Δ, E(T)′ 7Δ. Comitant.
  Distance (Worth 4-dot test and red glass): fusion and uncrossed diplopia.
  Fusion at near. 40 sec of arc steroacuity (Titmus).
  Excellent fusional amplitudes.
  Medical and neurologic examination normal.

  PAT (1½ hours in office):
  (cc) ET 30Δ, ET′ 27Δ.
  Because the patient was reluctant to consider surgery, a 15Δ base-out Fresnel prism was prescribed over his present glasses for a total prism correction of 27Δ.

  Follow-up:
  Two months: not comfortable with Fresnel.

  PAT (3½ hours in office):
  ET 52Δ, ET′ 60Δ.
  NB: Normal pupils and binocular visual acuity 20/20.
  All prisms removed.

  Surgery under microscope:
  Recession RMR 5 mm.
  Resection RLR 7 mm.

Follow-up:
Two and one-half weeks: (cc) E 2Δ, E′ 5Δ.
Two years:
>ACPT (cc) E 3Δ to 5Δ, E′ 4Δ.
>Distance and near: excellent sensory state.
>Asymptomatic.

Comments: This case illustrates the amount of deviation that can be found with prism and alternate cover test and the PAT. It shows also a patient with bifoveal fusion and a prism "build-up" response ("eats up the prisms"). It is not surprising that small amounts of prism could not help him. There was also no indication for bifocals because he had a normal near point of accommodation and a normal AC/A ratio. The excellent fusional amplitudes could have been the result of orthoptic exercises or from his own long-term effort to keep his deviation in check. This patient was a surgical candidate and could have been relieved of his symptoms many years before he was eventually cured if a PAT had been performed and the appropriate therapeutic conclusions had been drawn.

Heterophoria and the migraine syndrome. The symptomatic heterophoric patient often complains of long-standing, variable headaches, pain "in the back of the eyes," pulling sensation of the eyes, reading difficulty that quickly makes reading impossible or induces sleep, and even dizziness and nausea, especially in a moving vehicle. The same patient may also have difficulty in evaluating the relative speed of oncoming or passing vehicles when driving. Photophobia can be present. These patients often have a family history of migraine as well as heterophoria as it is frequent in the general population. Usually their amplitude of fusion is as good, if not better, than the asymptomatic patient. If their heterophoria remains small, with PAT it is reasonable to conclude that they do not present a "masked tropia." Although some patients with migraine and motion sickness may increase their threshold of tolerance with a weak optical or prismatic correction,[13] most are better handled medically or by avoiding precipitating factors. Acquired nonfamilial migraine is an important neurologic sign that should be investigated without delay.

Inverse prism for heterophoria. For the treatment of esophoria some authors advocate incorporating in the patient's glasses a 1Δ to 2Δ base-in prism to request divergence, and especially for exophoria, a 3Δ to 4Δ base-out prism to request convergence.[46;68] Much larger powers of inverse prisms are used during orthoptic exercise sessions to develop fusional amplitude in divergence or convergence. It is doubtful that an incorporated inverse prism of 1Δ to 2Δ truly benefits a patient with a genuine symptomatic heterophoria. On the other hand, if much above these powers, toleration of inverse prisms for permanent wear is unlikely.

Prisms for Fusional Amplitude Training

Prisms for fusional amplitude are reserved for patients with normal retinal correspondence, poor amplitude of fusion, and asthenopia. These exercises are based on the appreciation of physiologic diplopia, while using vergences to improve motor fusion.

Convergence exercise. The patient looks at the blunt end of a pencil held at arm's length and concentrates on seeing one, at the same time seeing two lights from a single distant light source because of the stimulation of nasal retinal points. The appreciation of uncrossed diplopia ensures that there is no suppression (Fig. 5-1, A). While maintaining it single, the pencil is *slowly* brought closer to the nose. The two distant lights move farther apart as the pencil approaches the nose, verifying that the test is being correctly performed. If one distant light disappears or the pencil becomes doubled, the exercise is discontinued and repeated from the beginning. *When teaching the exercise the examiner must also ensure with a quick cover-uncover test that the patient remains phoric.* In cases bordering on heterotropia, two lights and one pencil may be recognized and reported when in reality the patient sees two lights and two pencils but is ignoring one of the two images of the pencil. This condition is that of pathologic, not physiologic, diplopia. The cover test will disclose a movement of the eyes to take fixation, demonstrating that the pencil is not fixed binocularly. When the patient has learned to properly perform this exercise, divergence exercise is taught.

Divergence exercise. The patient looks at a distant light source while holding at arm's length the blunt end of a pencil that is seen to be doubled. The pencil is seen in crossed diplopia because of the temporal retinal point stimulation (Fig. 5-1, B). The doubled pencil is brought *slowly* toward the nose. The absence of suppression is verified when the distant light source is maintained single as the double images continue to separate. This exercise is sometimes referred to as "framing."

A patient with poor convergence may perform divergence exercises easily, whereas a patient with good convergence may perform them poorly. If a patient has overall poor vergences the examiner repeats their estimation, ensuring the full attention. Conversely, a patient who easily performs both exercises on the first attempt, during which time the examiner verifies with a rapid cover-uncover test that he or she has remained phoric, probably has a normal amplitude of fusion and is not a candidate for vergence training.

Exercise schedule. The exercises are done at home or at work in one 10-minute or two 5-minute sessions. For convergence deficiency, usually within 6 weeks, the patient should be able to maintain seeing a pencil single up to a point 4 cm (1½ inches) from the tip of the nose while still seeing the two lights at distance. A short exercise calling on divergence should follow each convergence training session to avoid accommodation-convergence spasm. It should be done even in a case of convergence insufficiency or if the divergence exercise is achieved without effort. The schedule is the same for

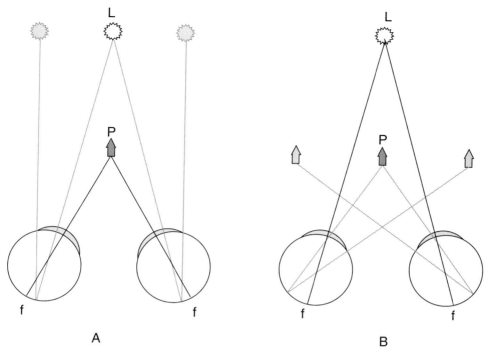

FIGURE 5-1. Physiologic diplopia. No suppression. **A,** Distant object is seen in uncrossed diplopia at distance when concentrating on near fixation object as a result of nasal retinal stimulation. **B,** Because of temporal retinal stimulation a near object is seen in crossed diplopia while fixating a distance object.

divergence insufficiency except that it is not followed by a convergence exercise.

In the absence of suppression the light source is replaced with an accommodative target. To stimulate interest, stereograms, based on the appreciation of physiologic diplopia, can be used as the test objects. In a symptomatic patient, prism exercises are prescribed to further improve fusional amplitudes.

Prism exercises. While the patient fixates on a target at ⅓ m base-out prisms of increasing power are placed sequentially in front of the dominant eye by using a bar or a Risley prism. To maintain a single image, the eyes must converge more as the strengths of the prisms increase. The patient confirms that the exercises are being correctly done by verifying physiologic diplopia, that is the blunt end of a pencil brought from the nose toward the target is seen doubled in crossed diplopia.

When used to elicit divergence, the exercise is done with base-in prisms. Both exercises are repeated at 6 m.

Jump convergence exercises. Jump convergence exercises are prescribed after convergence has been improved. With these exercises the patient must regain and maintain bifoveal single vision following a sudden change in convergence requirement. This is accomplished by quickly changing the fixation between a target at distance and a target at near or with prisms. In the prism exercise a base-out prism is held in front of one eye while the patient is asked to maintain fixation and fusion for a few seconds. The prism is then abruptly removed and replaced by increasingly stronger prisms until fusion can no longer be maintained. Once the patient has been taught the proper response, the exercise is carried out for 5- or 10-minute periods daily with weekly office control. A vertical or horizontal prism bar or both, as required, can be rented to the patient for use at home or at work.

A cooperative individual should normalize fusional vergences within 6 weeks. If by that time asthenopic symptoms are not improved, their etiology should be questioned, and a complete reevaluation should be carried out. It is an intriguing and sometimes a disturbing paradox to find an asymptomatic patient with very poor fusional amplitude or to have an asymptomatic subject rendered symptomatic just by performing these exercises for teaching or demonstration purposes.

Prism prescription for convergence insufficiency. Convergence insufficiency with normal or abnormal accommodation compensated by spectacles can be associated with a small phoria, an esophoria as well as an exophoria, and benefits mainly from orthoptic exercises. However, if there is an associated large exophoria in a presbyopic patient or in a patient with a neurologic condition such as myasthenia, the good amplitude of fusion developed through orthoptics may be maintained only with effort. In this case, incorporated base-in prisms are prescribed in the near correction after an office trial. The minimum amount that renders the patient comfortable is given.

If the patient has a symptomatic intermittent exotropia "convergence insufficiency type" (greater at near), the patient is managed according to the response to PAT (Case Presentation 5-3).

- **Case presentation 5-3**

  Patient G.W.: This 52-year-old women has asthenopic symptoms and intermittent horizontal diplopia for reading for 1 year. She has a past history of uveitis, presently inactive, and early onset of cataract.
  First exam:
  Visual acuity: OD +2.00 +2.50 × 20° = 20/25 −.
  OS +3.25 +1.75 × 165° = 20/20.
  Add +2.00 = J1 OU.
  Distance stereoacuity: 60 sec of arc (Vectograph).
  Near stereoacuity: 100 sec of arc (Titmus).
  ACPT: X 4Δ, X' 25Δ. Comitant.
  NPC: remote.
  Fusional amplitudes: D 8/6,   C 14/4.
  D' 16/14,   C' 16/12.
  PAT (in office):
  XT 12Δ, XT' 30Δ.
  Convergence exercise prescribed.

Follow up:
Six weeks: has done exercises. Same symptoms.
    Stereoacuity: 200 sec of arc (Titmus).
    NPC: remote.
    Fusional amplitudes: D 10/8,   C 14/4.
                            D' 25/20, C' 2.

Prism prescription:
    A Fresnel membrane 15Δ base-in, maximum power that does not induce diplopia at distance, is applied to the right lens of her bifocals (over eye with less vision OD).

Follow-up:
Five weeks: no diplopia, asymptomatic.
    Visual acuity: OD (ccΔ) = 20/70, (cc) = 20/30.
    After office trial, prism reduced to 5Δ base-in for each eye and incorporated in bifocal glasses.
One year: asymptomatic. Motility unchanged.
    Near stereoacuity: near 50 sec of arc (Titmus).

Comments: This case is not unusual. Taking into account the exodeviation of this patient at near, where she was symptomatic, she was able to converge about 40Δ at her first visit. It is not surprising that in her age group she could not develop a greater fusional amplitude, even with exercises, to acquire the extra 10Δ needed to become asymptomatic. Although she would be classified as an intermittent exotropia, convergence insufficiency type, she does not have a true convergence insufficiency (sensory problem), rather she has a motor problem. The prescribed prism acted on the motor condition by reducing the deviation to 15Δ, improving the fusional reserve to the point where the needed convergence is achieved without effort.

---------------------- Δ ----------------------

Prisms for Heterotropia

Presurgical prisms. Ophthalmology, as well as other surgical specialties, has benefited from the availability of smaller suture material and better magnification. Strabismus surgical techniques done under the microscope with direct visualization are more precise and thus easier to perform correctly.[78;81] Nevertheless, in strabismus surgery, even more than with other ophthalmic surgical procedures, proper surgical indications resulting from thorough evaluation and experience are essential to an optimal result.

Two goals for surgery must be considered: (1) sensory or functional to establish or restore binocularity, in addition to parallelism and (2) motor only without the hope of normalizing the binocular function (cosmetic surgery)

When sensory function is the primary intent of the surgery, it is indicated in some patients to normalize the sensory state with prismotherapy before surgery. Moreover, at the time of surgery, the full angle of deviation to be operated on is determined by PAT. In "motor" surgery, where the primary intent is to correct the apparent cosmetic defect, prisms are used only to ascertain if diplopia will be present or absent when the desired correction is achieved, not to determine the total (basic) surgical angle.

Early onset esotropia with the congenital esotropia syndrome. Truly congenital esotropia is rare. It has, together with constant esotropia of early onset, certain common characteristics.[17;30;43;51(pp293-305)] It is more often alternating with a large angle and can show simulated limited abduction, elevation in adduction, as well as manifest and latent nystagmus. Dissociated vertical deviation (DVD), patterns such as A and V, and vertical elements may be found during its evolution. Although infrequent, an accommodative component can be present.

In the early seventies, following the work of Hubel and Weisel,[32;33] I used Fresnel prism membranes as an adjunct to early surgery, for children of ages 6 to 12 months, to achieve parallelism for stimulating better sensory results. Surprisingly, with a proper frame, a child as young as 7 months would tolerate the slightly overcorrecting bilateral prisms, even when in an oblique orientation.[74] However, by the early eighties I abandoned prismotherapy in this group of patients because it did not improve the sensory results. Even if the eyes were straightened, by any means, before 1 year of age and if a careful follow-up had prevented the development of amblyopia, the long-term sensory result achieved in most cases was an alternating suppression. Only in rare cases were peripheral fusion and gross stereopsis achieved. To the present time the sensory outcome has remained about the same when parallelism has been restored between 6 and 24 months.[30;63] It is not yet established if the sensory outcome would be different if the eyes were straightened within the first 3 months of life. Specially designed prisms of a quality superior to those currently available could prove advantageous for such an investigation.

Pigassou-Albouy[61] is among the few who continue to use prismotherapy and other orthoptic methods to treat the deep sensory anomalies associated with this syndrome. She follows a treatment similar to the one she described for esotropia with ARC (see Treatment of ARC in this Chapter).

Small angle esotropia. The term small-angle esotropia includes very small-angle deviations that escape detection with a cover test but are revealed with the 4Δ test (see Chapter 3) up to angles of about 10Δ. In many of these patients, the common finding is that without any prismatic correction, they fuse the Worth 4-dot test and the Bagolini lenses (with or without a suppression area). This group has harmonious ARC (the angle of anomaly equals the objective angle), some amplitude of fusion, and a reduced stereoacuity. The acuity in the amblyopic eye varies from 20/100 to a negligible difference from the normal fellow eye affecting only the quality of vision. Other patients do not demonstrate any binocular link. In addition, some patients with accommodative esotropia, thought to be fully accommodative but who have a poor binocularity, will be found to have a microtropia when carefully examined.[34(p186);43;51(p313);55]

In general, the response to any treatment to cure the sensory anomaly is poor because it is deeply rooted and thus probably of very early onset. Notwithstanding, cures have been reported either spontaneously or after the successful treatment of the amblyopia,[41] after treatment with direct and inverse prisms, or following surgery.

In particular, in small-angle esotropia the direct prisms habitually provoke a marked angle buildup, that is, they are "eaten up." Even when operated for the built-up angle, the eyes often return to the original motor and sensory state or, more rarely, become exotropic. Except for well-supervised treatment of amblyopia in the child (see Case Presentation 5-1), in general these patients are better left alone unless decompensated. Change in eye dominance induced by an optical correction or with occlusion of one eye or its decreased vision from cataract or other pathology or with prismotherapy, or increased accommodation requirement with corrected progressive myopia may all trigger a decompensation. The goal of treatment is to return these patients to their original eye dominance and motor condition to allow them to function within their sensory anomalies.

Acquired nonparalytic esotropia in the child. Accommodative esotropia, which can appear as early as 2 to 3 months of age, responds well to patching and glasses, frequently achieving normal binocularity, in some cases even when the treatment is delayed. The deviation, because it is usually intermittent at its onset, allows the early development of a binocular link that can be later reestablished by treatment. For this reason it is important to determine through a questionnaire, documented by photographs, the true age of onset and the evolution of an esotropia. *For an acquired esotropia in the younger child, all measures should be taken to establish diagnosis and restore parallelism as soon as possible.* In the older child without a family history of strabismus, even in the presence of hypermetropia and full versions, a central origin has to be suspected. A normal ophthalmologic examination does not necessarily rule out the latter. If the esotropia is nonaccommodative, a full neurologic work-up, including magnetic resonance imaging (MRI), is advisable.[2;12;83]

Before considering surgery for the nonaccommodative element, in addition to having the full hypermetropic correction worn for a minimum of 3 months, with bifocals if indicated, it is wise to give a quick trial of prismotherapy to correct any residual esotropia or associated vertical that may prevent establishment of fusion. Some authors do not use prisms if the acquired esotropia presents with a vertical element. Other authors[4;47] and I will give oblique prisms for a therapeutic trial. *Membranes of powers above 20Δ are not used over both eyes* because of their potential detrimental effect on binocularity (see Chapter 2).

Technique for prismotherapy trial. If the deviation differs at near and distance the prismatic correction worn will be that for the larger deviation. If at any time within 1 hour "orthotropia" has not been observed and no stable correction can be found because of an angle buildup, the test is interrupted and the patient is treated according to the cosmetic blemish found *before PAT.* If "orthotropia" is quickly achieved and maintained, and the informed parents agree to prismotherapy, a trial is begun. Usually the minimum total power that achieves a control at near and distance is divided equally between the two eyes. A stronger power can be applied to the dominant eye as an antisuppression treatment. This technique is also used with a mild residual amblyopia. Although it is preferable to treat amblyopia before starting prismotherapy, a child may more readily accept wearing a prism membrane rather than a patch. The membrane should be supplemented by intermittent total occlusion is the visual acuity of the amblyopic eye fails to improve or begins to decline.

If sensory testing can be done and there is no improvement after 3 months of prismotherapy, it is discontinued. If sensory testing cannot be done, the response to the cover test is used to ascertain the presence or absence of fusion. If the deviation remains controlled as the prism power is decreased gradually, surgery can become unnecessary. Such good results have been reported by Bérard only in 10% of his cases.[80] In most cases, surgery has to be performed on the nonaccommodative component either for a sensory or for a cosmetic reason.

In my experience, *prismotherapy is much less effective in normalizing the sensory state in acquired esodeviation* (excluding consecutive esotropia) *than in exodeviation.*

It should be emphasized that patients who demonstrate ARC or an angle buildup do not receive prismotherapy. Patients are operated on the built-up angle found with PAT only if NRC can be elicited (Case Presentations 5-5 and 5-6).

- **Case presentation 5-4**

  Patient S.R.: This child of 3 years and 10 months had a history of acquired alternating esotropia, at about age 1 (confirmed with photos). She had undercorrected hyperopia and poor compliance with glasses (+2.00 OU).
  First exam:
  Visual acuity (cc): OD 20/30, picture chart (linear).
                      OS 20/40+.
  ACPT: (cc) ET 40Δ, ET′ 40Δ.
          (sc) ET 40Δ, ET′ 50Δ.
  Cycloplegic refraction: OD +5.50
                          OS +6.00
  For acceptance the correction was increased gradually to the maximum tolerated OD +4.75 and OS +5.25 and worn for 7 months.
  ACPT (cc): ET 20Δ, ET′ 25Δ.
  PAT (in office):
  ET 30Δ. Angle stable. Alternate suppression with all tests.
  Prismotherapy:
  Attempted because of history and response to PAT.
  15Δ base-out Fresnel membrane applied to each lens.
  Follow-up:
  Two months: excellent cooperation.
  Cover test: flick X to ⊕.

Worth 4-dot test: fusion up to ½ m.
Five months: (ccΔ) XT 6Δ, flick XT'.
   Sensory unchanged. Prism reduced OS 12Δ (total 27Δ).
   Red glass: alternated from crossed to uncrossed diplopia.
Eleven months (age 6):
   ACPT (ccΔ): flick ET, XT'.
   Sensory unchanged.
   Visual acuity (cc): OD = 20/25, OS = 20/25.
   Prisms removed. Parents happy with cosmetic appearance.

Comments: It is probable that, if this patient had received and worn her full hypermetropic correction at the onset of her esotropia at age 1, she would have demonstrated either a fully accommodative esotropia with a good binocular state or a microtropia with peripheral fusion and gross stereopsis. The period of neglect for almost 3 years at an age critical for the development of binocularity may explain the poor results. On the other hand, she was a good candidate for prismotherapy. She had worn her hypermetropic correction for more than 3 months, had equal vision, had a stable moderate angle, and at age 4 was willing and able to cooperate.

   This case illustrates my observation that prismotherapy rarely normalizes the sensory state in neglected constant esodeviations of childhood of onset as early as 1 year of age.

- **Case presentation 5-5**

Patient K.R.: This 3½-year-old girl had a history of right intermittent esotropia of sudden onset 3 months before, rendered alternating with patching. She would often close one eye or bump into objects. Her parents had observed a cyclic pattern of presentation as well as a variable esotropia. Her mother is myopic, and one cousin has esotropia.
First exam: constant esotropia
   Visual acuity (cc): OD = 20/30, picture chart (linear).
                              OS = 20/30.
   ACPT: ET 45Δ, ET' 45Δ. Concomitant.
   Cycloplegic refraction: +2.25 OU.
   Prescription: +2.00 OU (Maximum +) for 4 months.
   During that period the cyclic presentation could not be confirmed.
   Small accommodative component found.
   Neurologic work-up negative.
Presurgical examination:
   ACPT: (cc) ET 35Δ, A pattern 10Δ, ET' 35Δ to 40Δ. AC/A ratio:2/1
             (sc) ET 50Δ to 55Δ, ET' 50Δ to 55Δ.
PAT (in office):
   (cc) ET 50Δ, A pattern 10Δ, ET' 50Δ.
   Angle stable. NRC response.

Surgery under microscope:

Recession RMR 6 mm.

Recession LMR 6 mm.

An upward displacement of 4 mm was performed on each eye because of the presence of the A pattern with PAT at distance.

Follow-up:

Three weeks:

Worth 4-dot test: fusion at near and distance.

Vectograph: fusion, no stereopsis.

Titmus: 100 sec of arc stereoacuity.

No glasses, eyes straight, A pattern corrected.

One year: age 5 minus 2 months.

Stereoacuity: 60 sec of arc (Vectograph).

40 sec of arc (Titmus).

Two years:

Stereoacuity: 30 sec of arc (Vectograph).

40 sec of arc (Titmus).

ACPT (sc): ⊕, ⊕′.

Maddox rod: E 1Δ ⊖, X′ 1Δ ⊖′.

Normal amplitude of fusion.

Four years: Same sensory and motor results.

Comment: This child had an acquired, nonparalytic esotropia of sudden onset at age 3 years 2 months of unknown etiology. She responded very well to surgery performed 7 months later according to PAT. The buildup was taken into account because the patient had NRC. The 4 months of follow-up from the initial consult to surgery was to rule out an accommodative esotropia. Prismotherapy was not considered because it would have involved the use of high-power prisms over both eyes. The child also had no suppression. In such cases I favor the surgical restoration of parallelism as soon as the evaluation is completed. Also note that once the eyes were straightened, the binocular state stabilized and stereoacuity continued to improve without exercises as observed in other children without a history of strabismus.[66] By the age of 6 she demonstrated perfect stereopsis at both distance and near.

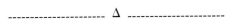

Residual esotropia. Residual esotropia is not infrequent following surgery. A patient with an acquired, nonaccommodative esotropia of an angle too large for a preoperative trial of prismotherapy may become a good trial candidate after the angle has been reduced surgically. First the effect of the hypermetropic correction has to be reevaluated. Often a deviation thought before surgery to be nonaccommodative will demonstrate an accommodative component after surgery. If not, the steps described in Technique for Prismotherapy Trial are followed.

Intermittent esotropia. Intermittent esotropia is often seen at the onset of accommodative esotropia and responds well to the wearing of a hypermetropic correction. It can also be seen in nonaccommodative esotropia. The trigger for the manifest deviation is not always easy to pinpoint. It resembles the one described for decompensated heterophoria, for example, the onset of myopia with the increased demand for accommodation and convergence when wearing the myopic correction. The clinical picture is similar to the one described for a masked tropia. Often long periods of control during childhood have allowed the development of a normal sensory state. Contrary to the habitual finding with intermittent exotropia, such patients will report blurred or double vision when the deviation becomes manifest. They may also have a reasonably good amplitude of fusion. A careful examination allows the differential diagnosis of a decompensated microstrabismus with ARC.

With a PAT, as with esophoria with a masked tropia, the angle of deviation is found to be large both at near and distance. These patients are not candidates for prisms, because to be effective the correction required would be too large, nor are they candidates for exercises. Because their binocularity is not deeply disturbed, once parallelism is restored, it spontaneously improves (Case Presentation 5-6).

- **Case presentation 5-6**

Patient M.K.: This 27-year-old woman had asthenopic symptoms. She could read only for short periods, got headaches, and "fell asleep." She had a history of intermittent esotropia since age 5, treated successively with optometric exercises, glasses for reading only, and bifocals around age 10. She had received prisms with "different orientations" since age 20, with temporary relief of symptoms.

First exam:
　　Visual acuity (pc):　OD −2.25　2Δ base-out = 20/20.
　　　　　　　　　　　　OS −2.50　+0.50 × 95°　2Δ base-out = 20/20.
　　Vectograph: fusion, no stereopsis. Titmus: 3000 sec of arc.
　　Worth 4-dot test: uncrossed diplopia, NRC.
　　ACPT (cc): E(T) 20Δ, E(T)′ 30Δ. Comitant.

PAT (in office):
　　ET 45Δ, ET′ 45Δ.
　　Not a candidate for prismotherapy.

Surgery under microscope:
　　Recession LMR 5 mm adjusted to 7 mm.
　　Resection RLR 7 mm.

Follow-up:
One week: flick E, ⊕′.
　　Stereoacuity (Vectograph): 60 sec of arc, (Titmus): 100 sec of arc.
　　Worth 4-dot test: fusion distance and near.
Six months: has noticed an "incredible difference." Asymptomatic.
　　ACPT (cc): E 4Δ, E′ 4Δ.

Stereoacuity (Vectograph): 60 sec of arc, (Titmus): 60 sec of arc.
Good fusion amplitudes.
Four years: status quo with contact lenses.
E 4Δ, E' 6Δ.

Comments: This patient was not a candidate for direct or a small amount of inverse prism "to improve her divergence," nor was she a candidate for exercises. Her basic esodeviation was too large to be held in check by a normal divergence amplitude. Her symptoms probably became worse because of increased accommodation demand with the development and progression of myopia. She could have been relieved of her symptoms much earlier if surgery had been proposed.

Because the patient had NRC, the buildup of the esotropia with PAT was essential to the determination of the amount of surgery to be done. The necessity to increase the recession during suture adjustment and the good outcome further reinforces the usefulness of PAT and of augmented surgery in such cases.

Esotropia with abnormal retinal correspondence (ARC). Abnormal retinal correspondence remains a challenge to all those interested in restoring binocular vision. Up to now the numerous attempts at its normalization have achieved a low success rate even under ideal conditions of treatment. Though prismotherapy, the most recent of this long list, seemed at first promising, its long-term effects have been disappointing.[64;71]

The reported normalization of retinal correspondence after prolonged consecutive exotropia following surgery in formerly esotropic patients, treated unsuccessfully with different orthoptic methods before surgery,[35] induced some surgeons to deliberately overcorrect esotropic patients with ARC even if more surgery was required later to achieve parallelism.[20] Because prisms could more easily provide such an overcorrection and negated the need for a secondary surgical intervention, they were used to replace the surgical overcorrection approach.

Various techniques were used, either a large overcorrection[1;8;57] or a small overcorrection[3;10] combined with an exact prism correction when NRC was demonstrated at some time during the treatment. The management of ARC included other methods of treatment such as patching, penalization, localization in space, afterimages, and sectors.

Unless the patient had previously undergone surgery, high-power prisms were often needed to achieve the planned overcorrection, especially with angle buildup. In 1961, when Bagolini first published his technique, Fresnel prisms (wafers and membranes) were not yet available, and his treatment was confined to residual esotropia with ARC not normalized by orthoptics before surgery. When Fresnel prisms became available (wafer prisms in 1965 and membrane prisms in 1970-1971), facilitating the correction of larger angles, many similar techniques were described that included their use for esotropia before surgery. However, it was soon recognized that the earlier the onset of the eso-tropia and the deeper the abnormal retinal correspondence, the poorer the sensory results that could be achieved. With few exceptions, prismotherapy came to be applied only in acquired esotropia with lightly rooted ARC (ortho-tropia quickly achieved) and finally only to esotropia with NRC.

Pigassou-Albouy is among the few exceptions. She was an early advocate of overcorrecting prisms before surgery for esotropia with ARC, and she used wafer prisms for large-angle deviations. In 1973 she reported a 75% to 85% cure in 840 such patients treated before age 8, with 80% compliance from both the parents and the child.[57] However, in 1988 she reported that 70% of her patients abandoned prismotherapy when prisms of large powers were pre-scribed, and for large-angle deviations she recommended surgery before pris-motherapy.[61]

Pigassou's method uses overcorrecting prisms for the treatment of ARC for patients under age 8 with no amblyopia or cured amblyopia,[57] and it follows these steps:

1. The eye preference is changed.
2. The angle is overcorrected by 15Δ to 20Δ for many months. Wafer prisms are used for high powers.
3. When retinal correspondence is normalized in space with bifoveal fu-sion tests (stereoprojector), the exact prismatic correction is given.
4. When suppression is overcome (after a few weeks), if there is no de-crease in the angle of deviation, surgery is done.

In 1983 Garcia de Oteyza and Torrubia reported their results using this method with nonaccommodative esotropia with onset from birth to age 4, no vertical element, no alphabetic pattern (A, V, X, or Y), no previous orthoptic treatment or surgery, and 2 years of follow-up after surgery. With deeply rooted ARC, the median prism overcorrection was 20Δ, and the duration of the treat-ment 11 months There was little improvement after 7 months. They reported 38.3% cure (28 of 73 cases) before surgery. After surgery, eight other cases were spontaneously normalized (11%).[25]

Although they reported fewer successes than Pigassou-Albouy, these results are remarkable considering that esotropia with deeply rooted ARC is thought by most authors to be an incurable disease. Nevertheless, Garcia de Oteyza considers that the cooperation requested from the patient and his or her fam-ily and the rigorous follow-up are disproportionate to the results achieved, and he no longer treats this anomaly.[26]

Bérard, Bagolini, and Aust have also changed their approach. Their cur-rent views should be kept in mind when referring to their earlier publications. Bérard has abandoned the treatment of ARC for many years.[11] Bagolini now believes, "If there is ARC and AFM (Anomalous Fusional Movements) are quite strong normal binocular vision can never be achieved." For this reason, con-trary to his previous recommendations,[5] he will not treat ARC with small-angle esotropia, early onset esotropia, or ARC with afterimage testing. He fur-ther states, "Even if normal retinal correspondence could be restored normal fusional movements do not develop and micro strabismus is usually the re-sult."[6;7] Aust still operates late when the child is 4 to 5, and uses alternate occlusion and prismostherapy as described in his earlier publications. However, what he formerly considered to be normalization of ARC, he now believes is in fact the persistence of a very small angle of anomaly that can only be detected with refined testing.[4]

Jampolsky has coined the term *prism orthoptics* with the goal of creating diplopia as an antisuppression and anti-ARC regime by the wearing of prisms during the immediate postoperative period of sensory motor plasticity.[38] He now may do prismotherapy if, with repeated PAT, the overconvergence response results in an ET of 5Δ to 8Δ. The prism power is reduced by 2Δ every 6 weeks to 2 months while maintaining the same small ET. After many months of gradual decrease the patient will have a small-angle ET without any prism. Therefore according to him, the prismotherapy has changed the basic deviation. Jampolsky has observed many such cases but concedes that the treatment is expensive and tedious and requires the full cooperation of the patient. His observation has allowed him to confirm the opinion of Haldén that there is "co-variation of the subjective and objective angles in ARC." In other words, ARC has adjusted to each succeeding smaller angle ET. When, and only when, a small-angle esotropia could be maintained at all times has Jampolsky observed a gradual decrease in the prism correction to the final result of a small-angle esotropia with no prisms.[39]

Comments and conclusions. There now seems to be general agreement that a small angle of anomaly, microtropia, peripheral fusion, and, at times, gross stereopsis are the best end results one can hope for in an esotropia with ARC, especially when deeply rooted, no matter what sequence of treatment has been followed. That such results can be observed after early surgery alone has also been concluded by a number of authors.[51;55;85] In the recent literature, these end results have been called a "cure" by some authors to justify their intensive treatment before surgery.[19;82] This approach not only delays the establishment of "orthotropia" but also puts an unnecessary burden on the child and the parents. There is also the risk to induce a permanent diplopia with the breakdown of suppression.

Consecutive esotropia. A consecutive esotropia following a surgically overcorrected intermittent exotropia is an excellent indication for prismotherapy, especially if the child is under 5 years of age. The younger the child, the quicker the suppression and the sooner an amblyopia and ARC will become established. Prisms are the treatment of choice if the child is emmetropic or myopic, if the esotropia is present at distance as well as at near, and if the esotropia has not spontaneously decreased after a few weeks of observation and part-time patching of the nondeviating eye. A base-out or oblique prism to correct any associated vertical deviation, with the horizontal component adjusted to leave the patient esophoric, is applied over the dominant eye or divided equally in the absence of suppression. Unless a mechanical restriction has been created surgically, with time, the power of the prism can be reduced, and eventually it can be removed[28;47;74] (Case Presentation 5-7).

- **Case presentation 5-7**

  Patient J.M.R.: This 8-year-old myopic child with intermittent exotropia had undergone prismotherapy with which she had achieved excellent sensory results at distance. She had also an easier control of her deviation: A power of 30Δ, required at the beginning of her prismotherapy, could be reduced to 16Δ. Though this management was preferred to surgery by the parents prior to this visit they now wanted the child to be relieved from wearing corrective prisms and requested the alternative of a surgical correction.

  Preoperative measurements:
  Visual acuity: OD -1.50 $-0.50 \times 85° = 20/20$.
  OS -1.00 $-0.50 \times 85° = 20/20$.
  ACPT: (cc) XT 30Δ, X(T)' 18Δ, comitant.

  PAT (in office):
  XT 35Δ, XT' 20Δ, +3.00 XT' 40Δ.

  Surgery under microscope:
  Bilateral 7 mm recession of lateral recti.

  Follow-up:
  Six days: (cc) ET 14Δ, E' 8Δ.
  Diplopia at distance only, overcome with 20Δ base-out prism.
  Fresnel applied over OD (preferred eye).
  Maddox rod (cc+Δ): E 2Δ ⊖, X' 6Δ ⊖'.
  Three weeks: prism can be reduced to 10Δ.
  Three months: slight increase in myopia. Needs 10Δ base-out prism over each eye for control.
  One year: prismatic correction can be reduced to 5Δ base-out prism over each eye. Three months later this was incorporated with a slightly increased myopic correction.
  Two years and three months:
  OD -2.25 $-0.50 \times 105°$ 5Δ base-out (inc) $= 20/20$.
  OS -1.75 $-0.50 \times 75°$ 5Δ base-out (inc) $= 20/20$.
  Prism correction cannot be reduced.

  Next follow-up visit:
  Four years:
  Vectograph: fusion, no stereoacuity.
  Titmus: 40 sec of arc.
  Worth 4-dot test: fusion distance and near.
  ACPT: (ccΔ) X 2Δ, X' 4Δ.
  OD -3.25 $-0.50 \times 100° = 20/20$.
  OS -2.25 $-0.75 \times 70° = 20/20$.
  Maddox Rod (cc no Δ): E 8Δ ⊖, E' 2Δ ⊖', +3.00 X' 10Δ.
  Glasses changed. No prism prescribed.
  Five years:
  Vectograph: fusion, 60 sec of arc stereoacuity.
  Titmus: 40 sec of arc.
  ACPT: E 2Δ to 4Δ, E' 4Δ.

Six and a half years: wants contact lenses.
   Vectograph: fusion, 30 sec of arc stereoacuity.
   Titmus: 40 sec of arc.
   ACPT: $\oplus$, E′ 2Δ to 4Δ.
   OD −4.25 −0.50 × 100° = 20/20.
   OS −3.00 −0.75 × 80° = 20/20.
Nine and one half years: wears contact lenses.
   Sensory unchanged.
   D 8/6, D′ 6/4,   C 25/20, C′ 30/25.
   ACT: small X, X′.
   Maddox rod: X 6Δ $\ominus$, X′ 10Δ $\ominus$′.
   OD −4.50 −0.50 × 100° = 20/20.
   OS −3.75 −0.50 × 80° = 20/20.

Comments: The measurements before and after PAT were similar. Because the deviation was greater at distance and the AC/A ratio was high (repeated measurements), a weakening procedure for both lateral recti muscles was chosen. The amount of recession done with my technique was in accordance with the deviation.[78] An immediate small overcorrection will usually decrease within a few weeks postoperatively. In this case the patient required a long-term prism prescription to recover and to maintain a normal sensory state. A Fresnel membrane over her preferred eye was first prescribed. When the deviation became stable and small, equal prisms were incorporated in her glasses. The progression of the myopia was an important factor in her delayed recovery. In addition, a decrease of the AC/A ratio with age may also serve to explain her latest findings. Surgery to correct her consecutive esotropia with time would have induced a relapse of the exotropia.

---------------------- Δ ----------------------

Blind spot mechanism. A small overcorrection, desirable in the immediate postoperative period, may suddenly increase to an angle of 25Δ to 30Δ. The object viewed by the deviated eye then falls on the blind spot, and the patient does not see double.[73] Although such a mechanism of suppression has been questioned and its existence even denied,[54] it can be observed clinically and responds well to prismotherapy[28] (Case Presentation 5-8).

- **Case presentation 5-8**

   Patient R.S.: This 14-year-old girl had a history of progressive myopia and intermittent exotropia, with onset at around 3 years of age. She had worn her first myopic correction (−1.75 OU) at age 7. Her sensory tests were normal at near. She had received prismotherapy to normalize her sensory state at distance. With a minimum of 6Δ base-in incorporated in each lens of her spectacle correction, she had a satisfactory control of her deviation. She wanted contact lenses.

Presurgical exam:

Visual acuity (cc): OD -6.50 $-2.00 \times 165°$ = 20/20 6Δ base-in.

OS -6.50 $-2.00 \times 165°$ = 20/20 6Δ base-in.

Stereoacuity: (Vectograph) 60 sec of arc, (Titmus) 40 sec of arc.

Worth 4-dot test: fusion distance and near.

ACPT (cc noΔ): LX(T) 35Δ, X' 25Δ, AC/A ratio:6/1.

PAT (in office):

XT 40Δ, XT' 45Δ.

Surgery under microscope:

Recession LLR 7 mm.

Resection LMR 7 mm.

Follow-up:

Five days: intermittent diplopia.

(cc) LET 16Δ, LET' 14Δ.

Myopic correction removed for near, exercises in levoversion.

Twelve days: no diplopia.

Vectograph: suppression OS at distance.

Worth 4-dot test: uncrossed diplopia.

(cc) LET 25Δ, LE(T)' 14Δ.

Because of increase in esotropia and suppression, blind spot mechanism suspected.

Prismotherapy:

25Δ base-out Fresnel membrane OD.

Follow-up.

Six weeks: fusion at distance and near reestablished.

No prism for ½ hour PCT (cc): flick X, flick X'. *Prism removed.*

Three months: horizontal diplopia reappeared.

ACPT: E(T) 6Δ, E(T)' 8Δ to 10Δ.

PAT (in office):

Angle buildup to ET 15Δ.

15Δ base-out Fresnel membrane OD, minimum power to overcome diplopia.

Five months: full binocularity distance and near with prism that can be reduced to 10Δ. Two months later the patient removed prisms on her own.

Ten months: ACPT: E 4Δ, E' 10Δ.

Four years and five months: same presurgical myopic correction. Has been wearing contact lenses for 3 years. Normal binocularity and fusion amplitude.

ACPT: E 2Δ, E' 6Δ, AC/A: 4/1.

Maddox rod: E 6Δ to 8Δ, E' 8Δ.

Comments: The decrease of the consecutive esotropia of this patient was not gradual as expected. The sudden increase from 16Δ, with diplopia, to 25Δ, without diplopia, suggests that the blind spot was being used as a mechanism of suppression. Would the relapse of the intermittent esotropia have built up to a similar angle if not treated with prisms? This case also illustrates how easily binocularity can be perturbed even at age 15 and recovered with a prismatic correction that should be decreased step-wise.

The AC/A ratio before surgery had not been repeated with PAT. The surgical decision was based on the response to PAT alone. With PAT the exodeviation became slightly greater at near than at distance, favoring a recession-resection rather than a bilateral lateral rectus recession.

Primary exotropia

Intermittent exotropia. Whether to use prismotherapy for intermittent exotropia, X(T), or not is determined by the individual strabismologist's philosophy of its overall management.

X(T) before age 5. For children of up to age 5, I favor a nonsurgical approach for X(T) for the following reasons: A consecutive esotropia induced after that age has less serious consequences.[21] The sensory state is not compromised.[23] The follow-up is not more demanding. There may be less need for reoperation.[65]

If there is good control at near (the position of gaze more often used in the young child) and a rare manifestation of alternating exotropia at distance, the child needs to be observed only every 6 months. The examiner must carefully evaluate the extent of the control. When feasible, *sensory testing and binocular visual acuity are done first* even if they are not recorded in that order. Dissociation is avoided while observing the size of the optotype at which one eye will begin to deviate.

If there is satisfactory control but strong preference for one eye, even in the absence of amblyopia, *part-time patching of the dominant eye is prescribed as an antisuppression treatment.*

If the control is poor at near and at distance and part-time patching has not improved it, or if the deviation is already alternating, the *response to minus lenses is determined.* If good control is achieved and there is no family history of myopia, minus lenses (−1.50 to −3.00) are given. The control will improve first at near. Rarely are stronger corrections needed.[15] It is very important to choose a spectacle lens shape that will not permit the child to peek outside the frame (see Chapter 2).

When there is a family history of myopia, I ordinarily do not prescribe overcorrecting minus lenses for medical-legal reasons. By inducing accommodation, minus lenses may be thought by some to be responsible for the development of myopia. There are treatments that claim to delay the progression of myopia by decreasing the accommodation requirement.[14] In the absence of a family history of myopia, I have never observed a child developing myopia after wearing overcorrecting minus lenses. However, in the absence of a family history of myopia, if the informed parents object to the prescription of minus lenses or if the AC/A ratio is low, *prismotherapy* is proposed.

Needless to say, the child should receive an adequate optical correction and amblyopia treatment before beginning prismotherapy. Usually amblyopia associated with primary exotropia is mild and responds quickly to patching. A deep or recalcitrant amblyopia signals a secondary exotropia.

A prism trial is done in the office to find the *minimum* power necessary to achieve control of the exodeviation at distance. We have found that the over-correcting technique, exceeding the larger deviation by 10Δ, as described by Hardesty[28] and later abandoned,[29] to be poorly tolerated by the patient because of the diplopia induced. The prism is applied only over the preferred eye as a partial occluder, taking advantage of the induced blur, or it is divided equally if there is no eye preference. Often within 6 months the prismatic correction can be decreased while the deviation remains controlled with good sensory response at distance and near. If the power needed is 16Δ or less, it can be equally divided and incorporated into appropriate spectacle lenses with a cosmetically acceptable thickness (Case Presentation 5-9).

It is unusual that X(T) cannot be managed nonsurgically before age 5, although surgery may become necessary later. For this rare instance surgery will be done according to the angle determined with PAT.

- **Case presentation 5-9**

  Patient D.H.: The parents consulted for this 4-year-old child because she had failed her "eye test" at school and because she was told her eyes "looked funny." There was a family history of myopia.
  First exam:
  Visual acuity (sc): OD = 20/40.
  OS = 20/50.
  ACPT (sc): XT 25Δ, X′ 8Δ. Comitant.
  Distance: alternate suppression with all tests.
  Near stereoacuity: 60 sec of arc (Titmus).
  OD −0.50 = 20/30 E chart (linear).
  OS −1.00 = 20/30 E chart (linear).
  With myopic correction for 5 months, no change in sensory state.
  ACPT (cc): X(T) 20Δ, flick X′.
  Prismotherapy:
  10Δ base-in Fresnel membrane, each eye.
  The treatment was well accepted by parents who wanted condition to be managed nonsurgically. Her cousin, also a patient, had been successfully treated with prisms.
  Follow-up:
  Three months: fusion at distance with Vectograph slide. Fresnel changed to
  5Δ base-in OS (nonpreferred eye): Total prismatic correction: 15Δ.
  Six months: the minimum prism power needed to maintain fusion at
  distance 10Δ is incorporated:
  OD −1.00 5Δ base-in (inc) = 20/25.
  OS −1.50 5 base-in (inc) = 20/25.
  Ten months: first indication of stereoacuity at distance: 120 sec of arc,
  near: 40 sec of arc.

Cover-uncover test (ccΔ): orthophoric. Likes her glasses.

One year and six months: incorporated prisms reduced to a total of 6Δ base-in.

Stereoacuity: distance 120 sec of arc, near 40 sec of arc.

Three years: full binocularity.

Incorporated prisms reduced to a total of 4Δ.

Five years and two months: prescription

OD −2.75 −0.50 × 145° 2Δ base-in (inc) = 20/20.

OS −2.50 −0.50 × 45° 2Δ base-in (inc) = 20/20.

Vectograph (cc): stereoacuity 60 sec of arc.

Titmus (cc): 40 sec of arc.

Comments: The power of the prisms prescribed was always the minimum needed to maintain a control of the X(T) at distance and the sensory state achieved. As can be seen in the last two follow-ups, a total of 4Δ was sufficient but essential to maintain the good result. It is probable that, by the time the patient wants contact lenses, the prismatic correction will have been removed. In this case there is no fear that the patient may lose the base-in prism effect of her spectacle correction for near (see Chapter 1) because the control of her deviation was always excellent at near even before her myopic correction or prismotherapy.

---------------------- Δ ----------------------

X(T) age 5 to 15. With a family history of myopia, a well-tolerated incorporated prism, and a good binocular state at near and distance there is little indication to rush into surgery in the "borderline" myope that has reached age 5. With the onset and progression of myopia the prismatic correction may often be discontinued. Myopes are more comfortable when exophoric than when esophoric. An esophoria may become decompensated with the progression of myopia.

If the child is hyperopic, with or without a family history of myopia, surgery is done when there is no amblyopia and when some fusion and stereopsis can be elicited at distance and near with the deviation corrected by clear conventional prisms (for a deviation of less than 24Δ) or Fresnel trial set prisms (for a deviation of greater than 24Δ) divided equally over the two eyes. In some children the binocular state at distance is more easily ascertained using an amblyoscope. The type and the amount of surgery to be performed is dictated by the response to the PAT.

In the presence of suppression the approach is the same as before age 5, *part-time patching and prisms*. Again, the minimum power of prism that achieves control is applied after trial in the office and favors the suppressing eye. An attempt to normalize binocular vision before surgery is indicated in this age group, especially if the exotropia has relapsed after a previous surgery. Although previously advocated,[76] prismotherapy is now not considered for abnormal retinal correspondence except when the patient has a dual retinal correspondence (ARC when the deviation is manifest, NRC when the deviation is controlled) and then with a guarded prognosis. According to the experience of Bagolini, in the presence of dual retinal correspondence with afterimages, it is very difficult to regain normal binocular vision for distance.[7]

In the presence of NRC the parents are informed that in successful cases a minimum of 6 weeks of prismotherapy is needed to establish fusion at distance and about 9 months to establish stereopsis at distance. Therefore the first follow-up visit is at about 6 weeks and then at 2- to 3-month intervals thereafter.

Even in this age group suppression can reappear quickly. When fusion and stereopsis have been established at distance, some patients will need only a small power prism (membrane or incorporated) to hold the deviation in check and to preserve the sensory results. They will revert to their previous suppression even if a power as little as 5Δ is removed.

In intermittent exotropia, with an angle of deviation of 20Δ or less, a complete cure with orthoptics, prisms, or minus lenses, used singly or in combination, has been reported.[15;18;62]

At the present time I favor prismotherapy for this group of patients as well and reserve orthoptic treatment mainly for patients with poor amplitude of fusion or with convergence insufficiency. I do not attempt to teach diplopia to the patient when the deviation is manifest.

Good binocular vision should never be presented as a guarantee against a relapse of an exodeviation, although it permits a better surgical prognosis. On the other hand, it is rare that an intermittent exotropia that has relapsed is not accompanied by a deteriorated binocular state (Case Presentation 5-10).

- **Case presentation 5-10**

  Patient J.I.: This 6½-year-old child had intermittent left exotropia, occasional diplopia, and headaches with a past history of V-pattern X(T), onset at age 2, bilateral lateral rectus recessions at age 2, and bilateral inferior oblique recessions 1 year later.
  First exam:
      Visual acuity OD = 20/20.
                  OS = 20/25.
      Distance (Worth and Vectograph): suppression OS.
      Near stereoacuity: 40 sec of arc (Titmus).
      ACPT: (sc) LXT 25Δ to 30Δ, flick X'. Comitant.
      Refraction (Cyclopentolate hydrochloride 1%):
      OD +0.75 +0.50 × 90°.
      OS +0.75 +0.50 × 90°.
      Treatment: part-time patching OD and minus lenses (-1.50 OU) for 4½ months equalized visual acuity but did not improve control at distance or eliminate suppression.
  Prismotherapy:
      25Δ base-in Fresnel over dominant eye.

Follow-up:
Three months: prism reduced to 15Δ base-in because of improved control
of X(T).
Six months: prism reduced to 10Δ base-in. First indication of fusion at
distance.
Nine months: first indication of stereoacuity at distance.
One year: slight deterioration of control at distance. Onset of myopia.
Prism was increased to 16Δ base-in and incorporated in his full myopic
correction.
OD −0.50 8Δ base-in (inc) = 20/20.
OS −1.00 8Δ base-in (inc) = 20/20.
Two years: (ccΔ) X 8Δ, X' 8Δ.
Distance stereoacuity: 120 sec of arc (Vectograph).
Near stereoacuity: 40 sec of arc (Titmus).

Comments: The child is asymptomatic. Because of the well-fit, cosmetically
acceptable incorporated prism correction, the parents do not wish any
further surgery, although they will reconsider it when the child is older and
may want contact lenses. The well-informed myopic father had no objection
to the myopic overcorrection prescribed for his son at the beginning of
treatment.

Adult patient. In the adult patient with intermittent or *apparently constant
primary exotropia*, prisms are used to determine the sensory state as well as
the indicated surgery. No attempt is made to normalize the sensory state with
prisms or with other nonsurgical methods. In rare cases, such as the patient
who cannot relax his or her convergence, a prolonged PAT (at home) becomes
necessary to determine the safe angle on which to operate. *Patients with constant large-angle primary exotropia of many years' duration may still demonstrate fusion when corrected with prisms*. Not only is this observation useful
regarding the improved surgical prognosis, but also it documents that the exotropia, intermittent at the onset, may become constant without definitively
altering the binocular state. After surgery in some patients, many months of
restored parallelism may be necessary before a normal binocular state is reestablished.

PAT for primary exotropia before surgery (patients with NRC). Surgery
should not be performed without a PAT, especially when the near and distance
measurements are very different.
The amount of prism power necessary to correct the greater deviation is
applied, and the test is performed as described (see Chapter 4). Often the examiner will observe that a patient with a larger deviation at distance *gradually* builds up the angle at near. Within 1 hour the deviation at near becomes

equal to, or sometimes even greater than, the distance deviation that may also have increased.

This response is different from:

1. That achieved when taking measurements with +3.00 lenses, relaxing the accommodation requirement. In fact, for a better AC/A ratio calculation, the latter is repeated with the prisms in place after the full angle of deviation has been found with PAT.
2. That obtained after prolonged occlusion of the nonpreferred eye, disrupting fusion.

With the total prismatic correction and time, the effort to fuse is neutralized without disrupting fusion. As for heterophoria, the true masked deviation is revealed with PAT. In the presence of a V or A pattern it is very important to repeat the measurements at distance with PAT in place. It is not uncommon to find either the disappearance of or a substantial decrease in the pattern, modifying the surgical plan.[77;79] A weakening procedure of the superior obliques is not recommended for an A pattern in a binocular patient. The reported consecutive overcorrection that resulted most probably was induced by surgery for a pseudo A pattern.

It has been suggested that "the unpredictability of surgery done on exotropic patients is due, in part, to there being no known best technique for determining the full deviation."[42] According to my experience, PAT is currently the most reliable test to reveal the full deviation as well as the muscle(s) to be operated on.[77;79] Other authors have confirmed its usefulness.[67]

If the deviation remains greater at distance after PAT, and this is rare, lateral rectus recessions are performed. This approach is also favored with a high AC/A ratio. Usually not only the AC/A ratio is normal, but the deviation at near will reach the distance deviation or slightly exceed it, and recession of one lateral rectus and a resection of one medial rectus are the procedures of choice. However, if the near deviation after PAT is greater than the distance deviation by 20Δ or more, resections of the medial recti are performed (Case Presentation 5-11).

For a large angle deviation, greater than 50Δ to 70Δ, a third muscle, a lateral or medial rectus, may have to be operated on. The choice of the muscle depends on the PAT response. For a very large angle, surgery can be done on both medial and both lateral rectus muscles. PAT will determine the type, recession or resection, and relative amounts of surgery on the respective muscles to correct an angle that is greater at distance gaze or greater at near gaze. *PAT influences the choice of the muscle to be operated on as well as the amount of the surgery,* especially when adjustable sutures cannot be employed.

- **Case presentation 5-11**

Patient S.H.M.: This 38-year-old woman had intermittent exotropia and difficulty performing close work. Diagnosed at age 14, she had been given bifocals and several series of optometric exercises, with only temporary relief.

First exam:

Visual acuity (sc) OD = 20/20.   Normal NPA.

OS = 20/20.

ACPT: (sc) X 20Δ, X′ 35Δ.

Distance: fusion only (Vectograph).

Near stereoacuity: 40 sec of arc (Titmus).

Excellent fusional amplitudes.

PAT (1 hour in office):

X 26Δ, X′(T) 46Δ.

Surgery under microscope:

Resection RMR 8 mm. Adjustable suture.

Resection LMR 8 mm. Adjustable suture.

On the following day, sutures were tied to leave patient slightly esophoric at distance. Diplopia was overcome with 8Δ base-out. Necessity of the immediate "minimal" overcorrection accepted by the anxious patient. No prism prescribed.

Follow-up:

Three weeks:

Distance stereoacuity: 30 sec of arc (Vectograph).

ACPT: X 2Δ to 4Δ, E′ 2Δ.

Six months: ACPT: X 4Δ, X′ 4Δ.

One year: condition unchanged.

Seen every 2 years to last follow-up at 11 years.

Eleven years: still asymptomatic with excellent binocularity.

(sc) X 4Δ, X′ 4Δ.

OD −0.50 × 90°   = 20/20 Add +1.50 = J1

OS −0.25 −0.50 × 90° = 20/20 Add +1.50 = J1

Comments: PAT further increased the difference between the near and distance deviations. It directed the choice of bilateral medial rectus resections rather than the recession of a lateral rectus and resection of a medial rectus. With time the latter procedure probably would have induced an overcorrection at distance or an undercorrection at near.

Sensory exotropia. With poor visual acuity in one eye, sensory exotropia is frequent. It is not an uncommon clinical observation that poor vision does not always protect against diplopia and its accompanying symptoms: nausea, loss of equilibrium, and disorientation. Although some patients can be immediately helped with a prismatic correction, others may require time to adjust (Case Presentation 5-13). Therefore, even with poor visual acuity in one eye, it is indicated to check with trial set prisms that diplopia will be absent once parallelism is achieved.

Conversely, with recovery of the visual acuity the control of the deviation habitually improves gradually, allowing the patient, if young enough, to develop some binocular link or later to return to an old anomaly (Case Presentation 5-12).

- **Case presentation 5-12**

Patient C.G.: The parents of this 12-year-old child consulted for a second opinion regarding surgery for a left exotropia. The symptoms were headaches and vertical diplopia. The patient did not wear glasses. At about age 8 her exotropia and amblyopia were discovered by a school nurse, which was confirmed by old photos. She was given glasses and patching, but the parents reported poor compliance. She had also worn contact lenses for 6 months during the year before this consultation.

First consultation:
Visual acuity: OD = 20/20.
OS = 20/400.
Refraction OS: $-3.00 -6.00 \times 7° = 20/30$.
No keratometric evidence of keratoconus.
ACPT: LXT 20Δ with diplopia, X′ 12Δ.
Near stereoacuity: 100 sec of arc (Titmus).
Worth 4-dot test: crossed diplopia.
At distance (cc): suppression in left eye with all tests.

PAT (in office):
(cc) LXT 24Δ, LHT 6Δ. No diplopia.
Aniseikonia: 4% to 5% (Awaya test)
Trial of contact lens for 6 weeks with intermittent diplopia.
Patient wants to try glasses before application of prisms.

Follow-up:
Two months: has been wearing glasses full time and prefers them to a contact lens on the left eye. Free of previous symptoms when not wearing glasses. Slight improvement in control, and sensory unchanged. Prism application deferred.
Lost to follow-up for 18 months. Has worn glasses faithfully. Does not see left eye turning out anymore. No diplopia.
Vectograph: fusion. No stereoacuity.
Titmus: 80 sec of arc stereoacuity.
Worth 4-dot test: fusion near and distance.
ACPT: X(T) 4Δ, X(T)′ 2Δ to 4Δ.
Bagolini: ARC harmonious.
Afterimages: ARC.

Comments: This patient had a decompensated microtropia. She benefited greatly from the correction of her anisometropia with a minus sphere and a high cylinder, which not only controlled with time her deviation but also improved her sensory function. In this age group, such a trial is always indicated before considering a prism prescription or surgery. This patient has an ideal refractive error for later excimer laser refractive surgery should she wish to be dispensed from glasses and is unable to wear a contact lens.

--------------------- Δ ---------------------

- **Case presentation 5-13**

Patient C.S.: This 19-year-old woman had constant, very annoying horizontal diplopia and exotropia with poor vision in the right eye. Her previous history revealed a traumatic cataract with iris prolapse of the right eye from a penetrating injury at age 7. A cataract extraction was performed at age 8, followed by many unsuccessful attempts at contact lenses wear for correction of her visual acuity.

First exam:

Visual acuity (pc): OD plano = CF at 1 m.

OS $-4.25 -0.25 \times 175° = 20/20$.

Refraction: OD $+11.00 -2.00 \times 165° = 20/25$.

ACPT: RXT 35Δ, RXT′ 35Δ. Flick hypotropia.

Fusion (red glass): crossed diplopia, NRC.

Prism perscription:

After prolonged office trial, a 30Δ Fresnel membrane, base at 3°, was applied to the plano right lens. Although the distressed patient still saw a peripheral rim around the now superimposed diplopia images, she was encouraged to try the membrane.

Follow-up:

Intermittent diplopia for the first month. Then no diplopia with a follow-up of 5 years.

Comments: The Fresnel membrane was replaced on an annual basis over the plano lens. The patient did not want a balance minus lens that would have improved cosmetic appearance. Diplopia reappeared on any attempt to reduce the power of the prism although visual acuity of the right eye is only finger counting at 1 m. If the patient had accepted to have the iris prolapse repaired, muscle surgery with an adjustable suture would have been done, and her aphakia would have been corrected. She is happy with the Fresnel membrane and does not wish to have further intraocular surgery. She is attending law school without vision difficulty and is very grateful for having been relieved of her long-standing diplopia with the accompanying distressing symptoms.

--------------------- Δ ---------------------

Consecutive exotropia (NRC and ARC). After surgery for esotropia in the young child, consecutive exotropia, that is still evident after 1 month, is infrequently encountered. Consecutive exotropia more commonly occurs in the adolescent or in the adult. In some cases, especially when a binocular link is present and the exotropia is treated immediately, small base-in prisms incorporated in the minimum hyperopic correction or in the maximum myopic correction allowing best vision may succeed to hold the deviation in check or reestablish a microtropia. More often, prismotherapy alone has been found to be unsuccessful in securing stable fusion in surgically overcorrected esotropia.[37]

If the patient sees double and the full prism correction allows fusion or a return to the old pattern of suppression, surgery is done accordingly, If not, temporary prisms are given to overcome diplopia and, if sufficiently large, to determine the safe angle to be operated on. After an office trial, a single Fresnel membrane is applied over the lens of the nondominant eye. If the adult patient has ARC, the amount of prism correcting the *subjective* angle is given. Afterward a stronger prism is given, provided it does not provoke diplopia. The ARC response to testing may be that of ARC with esotropia that has not changed with the consecutive exotropia (see Fig. 3-12). No attempt is made to normalize the retinal correspondence. After a trial with a Fresnel membrane, if the angle is small, incorporated prisms are prescribed. If the angle is large enough to be cosmetically unsatisfactory and the patient will cooperate to the technique, surgery with an adjustable suture is the procedure of choice. The prism trial test will demonstrate the angle that needs to be corrected. Because it will not correspond to the objective angle, the patient will have to be informed of and accept the postoperative deviated eye position as well as the possibility of a relapse of diplopia necessitating a second intervention or a prismatic correction. During the suture adjustment it is necessary, while compensating the diplopia, to anticipate variations in eye position that occur with time. For example, unless a mechanical factor has been created, a resection procedure relaxes within a few weeks postoperatively. Therefore if diplopia is present, when adjusting the muscle it is wise to slightly overcorrect and to prescribe a temporary prism membrane postoperatively.

REFERENCES

1. Adelstein FE, Cüppers C: Le traitement de la correspondance rétienne anormale à l'aide des prismes, *Ann Oculist (Paris)* 203:445, 1970.
2. Anderson WD, Lubow M: Astrocytoma of the corpus callossum presenting with acute comitant esotropia. *Am J Ophthalmol* 69:594-598, 1970.
3. Aust W, Welge-Lussen L: Effect of prolonged use of prisms on the operative results of strabismus. *Ann Ophthalmol* 3:517-523, 1971.
4. Aust W: Personal communication, May 1993.
5. Bagolini B: Postsurgical treatment of convergent strabismus with a critical evaluation of various tests. In Schlossman A, Priestly B, editors: *Int Ophthalmol Clin*, vol 6, no 3, Boston, 1966, Little, Brown and Co, pp 633-667.
6. Bagolini B, Zanasi MR, Bolzani R: Surgical correction of convergent strabismus: its relationship to prism compensation. *Doc Ophthalmol* 62: 319,321, 1986.
7. Bagolini B: Personal communication, May 1993.
8. Baranowska-George T: Application of penalization and prismatic overcorrection in the treatment of strabismus by the localization method. In Fells P, editor: *The Second Congress of the International Strabismological Association, Marseille 1974,* Marseille, 1976, Diffusion Générale de Librairie, pp 229-240.
9. Bérard PV, Reydy R, Layec M, Berthon J: The permanent wearing

of prisms in heterophoria and ocular palsy. In Fells P, editor: *The Second Congress of the International Strabismological Association, Marseille 1974*, Marseille, 1976, Diffusion Générale de Librairie, pp 241-246.

10. Bérard PV, Reydy R., Berthon J: Permanent wearing of prisms and early delayed treatment of esotropia. In Moore S, Mein J, Stockbridge L, editors: *Orthoptics: past, present, future, transactions of the Third International Orthoptic Congress, Boston 1975*, New York, 1976, Stratton Intercontinental Medical Book Corporation, pp 203-208.

11. Bérard PV, Quéré E, Roth A, Spielmann A, Woillez M: *Chirugie des strabismes*. Paris, 1984, Masson. pp 243, 249.

12. Bixenman WW, Langanu JF: Acquired esotropia as initial manifestation of Arnold-Chiari malformation, *J Pediatr Ophthalmol Strabismus*, 24:83-86, 1987.

13. Black-Kelly TS: Arrest of migraine or headache with prisms, *Prac Med* 225:1504-1505, 1981.

14. Brodstein RS, Brodstein DE, Olson RJ, Hunt SC, Williams RR: The treatment of myopia with atropine and bifocals. *Ophthalmology* 91:1373-1379, 1984.

15. Caltrider N, Jampolsky A: Overcorrecting minus lens therapy for treatment of intermittent exotropia, *Ophthalmology* 90:1160-1165, 1983.

16. Carter DB: Fixation disparity and heterophoria following prolonged wearing of prisms, *Am J Optom Arch Am Acad Optom* 42:141-152, 1965.

17. Ciancia A: Early esotropia. In Ferrer O, editor: *Ocular motility: Int Ophthalmol Clin*, vol 2, no 4, Boston, 1971, Little, Brown and Co, pp 81-87.

18. Cooper EL, Leyman IA: The management of intermittent exotropia. A comparison of the results of surgical and nonsurgical treatment. In Moore S, Mein J, Stockbridge L, editors: *Orthoptics: past, present, future, transactions of the Third International Orthoptic Congress*, New York, 1976, Stratton Intercontinental Medical Book Corporation, pp 563-568.

19. Christenson GN: Treatment of esotropia with anomalous correspondence, *J Am Optom Assoc* 63:257-261, 1992.

20. de Decker W: Results of surgery versus prism tolerated overcorrection therapy of anomalous correspondence. In Fells P, editor: *The Second Congress of the International Strabismological Association, Marseille 1974*, Marseille, 1976, Diffusion Générale de Librairie, pp 279-282.

21. Edelman PM, Brown MH, Murphree AL, Wright KW: Consecutive esodeviation . . . Then what? *Am Orthop J* 38:111-116, 1988.

22. Fletcher MC, Abbot W, Girard LJ, Guber D, Silverman SJ, Tomlinson E: Results of biostatistical study of the management of suppression amblyopia by intensive pleoptics versus conventional patching, *Am Orthop J* 19:8-30, 1969.

23. Fletcher MC: Natural history of idiopathic strabismus. In Allen J, editor: *Symposium on strabismus: New Orleans Academy of Ophthalmology*, St Louis, 1971, Mosby, p 17.

24. Garcia de Oteyza J, Fernandez Agrafojo D, Cardozo Villanueva E, Rodriguez Alvarez F, Perosanz C: Resultados en el trátamiento de la fijacion excentrica con prismo inverso. In Murube J, editor: *Acta XVII Concili Europaeae Strabolologicae Associetatis*, Madrid, May 1988, pp 167-179.

25. Garcia de Oteyza JA, Torrubia R: Etude statistique du résultat du traitement prismatique, *J Fr d'Orthoptique* 15:71-78, 1983.

26. Garcia de Oteyza JA: Personal communication, Sept 1992.

27. Gillies WE: The use of prisms for the treatment of diplopia caused by decompensation of distance esodeviation in older patients. In Fells P, editor: *The Second Congress of the International Strabismological Association, Marseille 1974*, Marseille, 1976, Diffusion Générale de Librairie, pp 288-290.

28. Hardesty HH: Treatment of under and overcorrected intermittent exotropia with prism glasses, *Am Orthop J* 19:110-119, 1969.

29. Hardesty HH, Boynton JR, Keenan JP: Treatment of intermittent exotropia, *Arch Ophthalmol* 96:268-274, 1978.

30. Helveston E: 19th Annual Frank Costenbader Lecture—the origins of congenital esotropia, *J Pediatr Opthalmol Strabismus* 30:215-232, 1993.

31. Henson DB, North R: Adaptation to prism-induced heterophoria. *Am J Optom Physiol Optics* vol 57, no 3, pp 129-137, 1980.

32. Hubel DH, Weisel TN: Receptive fields of cells in striate cortex of kittens reared with artificial squint. *J Neurophysiol*, 26:994-1002, 1963.

33. Hubel DH, Weisel TN: Binocular interaction in striate cortex of kittens reared with artificial squint. *J Neurophysiol*, 28:1041-1059, 1965.

34. Hugonnier R: *Strabismes, hétérophories, paralysies oculomotrices—les désiquilibres oculo-moteurs en clinique.* Paris, 1959, Masson et Cie.

35. Hugonnier R: The influence of the overcorrection of an esotropia on abnormal correspondence. In Arruga A, editor: *International Strabismus Symposium, Giessen 1966,* Basel, 1968, S Karger, pp 307-310.

36. Hugonnier R, Clayette Hugonnier S: *Strabismus, heterophoria, ocular motor paralysis, clinical ocular muscle imbalance,* St Louis, 1969, Mosby, (translated and edited by Véronneau-Troutman S), pp 239-240.

37. Iacobucci I: The value of Press-on prisms in the management of partially accommodative esotropia and surgically overcorrected esotropia, *Am Orthop J* 27:91-96, 19-77.

38. Jampolsky A: The postoperative use of prisms. In Fells P, editor: *The Second Congress of the International Strabismological Association, Marseille 1974,* Marseille, 1976, Diffusion Générale de Librairie, pp 291-294.

39. Jampolsky A: Personal communication, June-July 1993.

40. Kara GA: Personal communication, July 1993.

41. Keiner ECJF: Spontaneous recovery in microstrabismus, *Ophthalmologica (Basel)* 177:280-283, 1978.

42. Kushner BJ, Lucchese NJ, Morton GV: The influence of axial length on the response of strabismus surgery, *Arch Ophthalmol* 107:1616-1618, 1989.

43. Lang J: Special forms of comitant strabismus. In Reineke R, editor: *Strabismus II, Proceedings of the Fourth International Strabismological Association, Asilomar, Calif, 1982,* New York, 1984, Grune & Stratton, pp 770-789.

44. Lie I, Opheim, A: Long-term acceptance of prisms by heterophorics, *J Am Optom Assoc* 56:272-278, 1985.

45. Lie I, Opheim A: Long-term stability of prism correction of heterophorics and heterotropics: a five year follow-up, *J Am Optom Assoc* 61:491-498, 1990.

46. Linksz A: Therapeutic uses of prisms: Discussion: In Ferrer O, editor, *Ocular motility: Int Ophthalmol Clin,* vol 11, no 4, Boston, 1971, Little, Brown and Co, pp 292-295.

47. Moore S, Stockbridge L: An evaluation of the use of Fresnel Press-On prisms in childhood strabismus, *Am Orthop J* 24:62-66, 1975.

48. Milder DG, Reineke RD: Phoria adaptation to prisms, *Arch Neurol* 40:339-342, 1983.

49. Nawratski I, Oliver M: Eccentric fixation managed with inverse prism, *Am J Ophthalmol* vol 71, no 2, pp 549-552, 1971.

50. von Noorden GK: Experiences with pleoptics in 58 patients with strabismic amblyopia, *Am J Ophthalmol* 58:41, 1964.

51. von Noorden GK: *Binocular vision and ocular motility: theory and management of strabismus,* ed 4, St Louis, 1990, Mosby.

52. North R, Henson DB: Adaptation to prism induced heterophoria in subjects with abnormal binocular vision

or asthenopia, *Am J Optom Physiol Optics* 58:746-752, 1981.

53. Ogle KN, Martens TG, Dyer JA: *Oculomotor imbalance in binocular vision and fixation disparity,* London, 1967, Henry Kimpton.

54. Olivier P, von Noorden GK: The blind spot syndrome. Does it exist? *J Pediatr Ophthalmol Strabismus* 18:20-22, 1981.

55. Parks MM: Monofixation syndrome. In Tasman W, editor, and Jaeger EA, asst editor: *Duane's clinical ophthalmology,* vol 1, Philadelphia, 1992, JB Lippencott. pp 1-12.

56. Pigassou-Albouy R, Garipuy J: Traitement de la fixation excentrique strabique par le port d'un prisme et l'occlusion, *Bull Mém Soc Fr Ophtalmol* 3:367-382, 1968.

57. Pigassou-Albouy R, Garipuy J: The use of overcorrecting prisms in the treatment of strabismic patients without amblyopia or with cured amblyopia, *Graefes Arch Clin Exp Ophthalmol* 186:209-226, 1973.

58. Pigassou-Albouy R: Treatment of abnormal retinal correspondence combined with eccentric fixation by calibrated inverse prisms and total occlusion. In Fells P, editor: *The Second Congress of the International Strabismological Association, Marseille 1974,* Marseille, 1976, Diffusion Générale de Librairie, pp 257-278.

59. Pigassou-Albouy R: Traitement du strabisme par les prismes. Mise au point 1984. *Doc Ophthalmol* 60:45-69, 1985.

60. Pigassou-Albouy R: A propos du traitement de la fixation excentrique, "Prisme inversé et occlusion," *J Fr Ophtalmol* 11:597-600, 1988.

61. Pigassou-Albouy R: Prism therapy for strabismus, *J Ophthal Nurs Technol* 7(1):18-25, 1988.

62. Pratt-Johnson JA, Tillson MB: Prismotherapy in intermittent exotropia: a preliminary report, *Can J Opthalmol* 14:243-245, 1979.

63. Pratt-Johnson JA: Fusion and suppression: development and loss, *J*

Pediatr Ophthalmol Strabismus 29:4-11, 1992.

64. Quéré MA: *Physiopathologie clinique de l'Équilibre oculomoteur. Études pratiques de sensorio-motricité oculaire.* Paris, 1983, Masson, p 192.

65. Richard JM, Parks MM: Intermittent exotropia: surgical results in different age groups, *Ophthalmology* 90:1172-1177, 1983.

66. Romano PE, Romano JA, Puklin JE: The development of normal binocular single vision in childhood. In Moore S, Mein J, Stockbridge L, editors: *Orthoptics, past, present, future: transactions of the Third International Orthoptic Congress, New York, 1976,* New York, 1976, Stratton Intercontinental Medical Book Corporation, pp 11-23.

67. Ron A, Merin S: The pre-op adaptation test (PAT) in the surgery of exotropia, *Am Orthop J* 38:107-110, 1988.

68. Rubin W: Reverse prism in ocular motility problems. In Ferrer O, editor: *Ocular motility: Int Ophthalmol Clin,* vol 11, no 4, Boston, 1971, Little, Brown and Co, pp 263-268, 292-295.

69. Rubin W: Reverse prism and calibrated occlusion in the treatment of small-angle deviations, *Am J Ophthalmol* 59:271, 1965.

70. Salvi G, Frosini R, Boschi MC, Galassi F: Possibilités de rééducation fonctionelle chez le microstrabisme avec fixation excentrique, *Bull Mém Soc Fr Ophtalmol* pp 158-164, 1976.

71. Stangler-Zuschrott E: Acht jahre prismenbehandlung des strabismus convergens alternans. (Eight years treatment of convergent alternating squint), *Klin Mbl Augenheik* 177:835-838, 1980.

72. Swan KC: Esotropia following occlusion, *Arch Ophthalmol* 37:444-451, 1947.

73. Swan, KC: The blind spot mechanism in strabismus. In Allen J, editor: *Strabismus Symposium II, New Orleans Academy of Ophthalmology,* St Louis, 1958, Mosby, pp 201-211.

74. Véronneau-Troutman S: Fresnel prism membrane in the treatment of strabismus, *Canad J Ophthal* 6:249-257, 1971.

75. Véronneau-Troutman S, Dayanoff S, Stohler T, Clahane A: Conventional occlusion versus pleoptics in the treatment of amblyopia, *Am J Ophthalmol* 78:117-120, 1974.

76. Véronneau-Troutman S, Shippman S, Clahane AC: Prisms as an orthoptic tool in the management of primary exotropia. In Moore S, Mein J, Stockbridge L, editors: *Orthoptics: past, present, future, transactions of the Third International Orthoptic Congress, Boston 1975*, New York, 1976, Stratton Intercontinental Medical Book Corporation, pp 193-201.

77. Véronneau-Troutman S: Practical principles, guidelines and philosophy of the management of strabismus—what to do and when. In Boyd B, editor: *Highlights of ophthalmology*. Silver Anniversary Edition 1981, vol 1, New York, 1981, Arcata Book Group and Tennessee, 1981, Kingsport Press, pp 476-485.

78. Véronneau-Troutman S: Strabismus microsurgery. In Jakobiec FA, Sigelman J: *Advanced techniques in ocular surgery*, Philadelphia, 1984, Saunders, pp 667-679.

79. Véronneau-Troutman S: Surgical implications of the prism adaptation test, *Binocular Vis* 1:107-109 1985.

80. Véronneau-Troutman S: Interview with Bérard P: "Acquired esotropia cases good prism candidates if seen early," Marseille, *Ophthalmology Times*, pp 25-26, Oct 15, 1986.

81. Véronneau-Troutman S: Prerequisites for strabismus microsurgery. Microscope, instruments and sutures. In Draeger J, Witmer R, editors: New microsurgical concepts II. Cornea, posterior segment, external microsurgery, Proceedings of the International Microsurgery Study Group, Funchal, Madeira, 1988, *Dev Ophthalmol* vol 18, Basel, 1989, S Karger, pp 24-35.

82. Wick B, Cook D: Management of anomalous correspondence: efficacy of therapy, *Am J Optom Physiol Optics* 64:405-410, 1987.

83. Williams AS, Hoyt CS: Acute comitant esotropia in children with brain tumors, *Arch Ophthalmol* 107:376-378, 1989.

84. Worrell BE Jr, Hirsch MJ, Morgan MW: An evaluation of prism applied by Sheard's criterion, *Am J Optom Arch Am Acad Optom* 48:373-376, 1971.

85. Wybar K: The use of prisms in pre-operative and post-operative treatment. In Fells P, editor: *The First Congress of the International Strabismological Association, Acapulco, Mexico, 1970*, London, 1971, Henry Kimpton, pp 249-255.

CHAPTER 6

Nystagmus

Nystagmus is a repetitive back and forth movement of the eye(s). It is termed *pendular* when the oscillations have about equal velocity and amplitude. It is termed *jerk* when a slower phase in one direction is followed by a quick corrective movement in the opposite direction. The two types often exist in the same patient. Although the slow phase usually reflects the underlying abnormality, *jerk nystagmus is qualified by the direction of the rapid component. If the latter is to the right the nystagmus is described as "right beating" or to the left as "left beating."*

Nystagmus can be monocular or binocular as well as horizontal, vertical, torsional, or any combination of these directions. Pathologic nystagmus can be either congenital or acquired. As a rule only patients with acquired nystagmus will complain of the *apparent movement of the environment (oscillopsia)*. Intoxication and lesions along the vestibular, oculogyric, and optic pathways can provoke a nystagmus that becomes part of other otoneuroophthalmic signs. Nystagmus can be temporarily induced in normal individuals in response to vestibular and optokinetic stimuli and can therefore be used to test the integrity of these systems.

Manifest Congenital Nystagmus

Although termed *congenital,* this nystagmus is rarely seen at birth and its onset is usually during the first 3 months of life. Nystagmus observed early in life may be associated with metabolic disease, structural abnormalities of the brain, and more rarely with tumor in the region of the optic chiasm and third ventricle (unilateral nystagmus and head nodding)[45(p893)] However, in most patients such etiology is absent. It is rarely familial unless associated with a hereditary condition. Its prevalence in the general population is reported to be 1 in 6550, with a slight preponderence of males.[17(p264-265)]

Sensory nystagmus. Some patients demonstrate a nystagmus secondary to a disturbance of the visual input, termed *sensory.* It is observed in congenital cataract, optic nerve hypoplasia, Leber's amaurosis, albinism, achromatopsia, aniridia, or other pathology that will markedly interfere with adequate image formation or transposition during initial vision development. If it can be cor-

145

rected early and the integrity of the optics can be restored, as in congenital cataract, the nystagmus may disappear. This has also been observed in aniridia patients fitted with a pinhole contact lens. Sensory nystagmus is more often pendular.

Motor nystagmus. Most patients with congenital nystagmus have no primary visual disturbance. Their nystagmus is termed *motor.* Although it has been postulated that a defect of the brainstem gaze-holding network may be the cause, the etiology is unknown. The decrease in visual acuity is considered secondary to the eye movements. However, the reduction in nystagmus intensity is not necessarily followed by a gain in visual acuity. Large recession procedures on the horizontal recti have been recommended to dampen the eye movements and improve acuity.[3;23] At best the adult gains one or two lines of vision.[18] Nevertheless, the gain in cosmetic appearance and in the ability to function visually may be much greater.[27(p451);28] The procedure also has been proposed in the child to prevent the development of bilateral amblyopia secondary to the eye movements that may have been responsible for the poor visual acuity results of this procedure in the adult. The interesting observation in some cases of nystagmus is that the movement stops with the occlusion of one eye, and the visual acuity of the uncovered eye improves (inverse latent nystagmus).[17(pp634-647)] This response may happen on occlusion of the eye with the poorer vision or of the eye with the better vision and, when the vision is equal, only with the right eye or only with the left eye.

In general, the vision is better in patients having a longer period of rest between the saccades *(foveation time)* and a low velocity of image motion during the slow phase.[45(p896)] Acuity frequently improves at near, allowing the child or young adult to attend regular school. About one third of the patients have a normal or "almost normal" vision at both near and distance.[16] Nystagmus decreases with age and in some instances disappears.[2]

Motor nystagmus is more often horizontal and of the jerk type. Either the slow or the fast phase can be toward the fixating eye. The nystagmus may dampen with a head turn, a chin depression or elevation, or a head tilt, alone or in combination. If the slow phase in each eye is in the same direction and amplitude, the nystagmus is said to be *concordant (conjugate);* that is, the patient will adopt always the same head posture to favor the gaze with least nystagmus *(null point, neutral zone),* no matter which eye is fixating. It can be *discordant (disconjugate),* as found in patients with a null point in right *and* left gaze; that is, the head posture reverses with the fixating eye. It is termed *bidirectional alternating* when the abnormal head posture reverses in monocular as well as in binocular fixation.

Nystagmography is particularly useful to identify those patients with manifest congenital nystagmus who have both pendular and jerk nystagmus. The association with strabismus is about the same as that found in the general population without nystagmus, and the covering of one eye does not increase the nystagmus, contrary to a superimposed *latent* or *manifest-latent* nystagmus.

Latent and Manifest-Latent Congenital Nystagmus

The nystagmus is said to be *latent* when it is absent with both eyes open and *manifest-latent* when it is present but damped. *The nystagmus is provoked or the beat is increased by occlusion.* Adduction decreases the saccades and improves vision. The fast phase is toward the fixating eye. Infantile or early onset esotropia (congenital esotropia syndrome)[6;21] is frequently associated with latent or manifest-latent nystagmus. For Dell'Osso the presence of strabismus is essential for the diagnosis, in contradistinction to manifest congenital nystagmus that is infrequently associated with esotropia.[13] With a cooperative patient, *nystagmography* will allow a clear-cut differential diagnosis. *The manifest nystagmus will demonstrate a slow phase with increasing velocity while the latent and manifest-latent have a slow phase with decreasing velocity.*

When examining a congenital nystagmus it is important not only to evaluate the effect of covering one eye but also to compare the visual acuity with an occluder, such as the Spielmann type that allows the observation of the eye under cover, and with a blurring device, such as a +6.00 or +8.00 lens. For better results the head of the patient should be in the position of least nystagmus during the refraction, vision evaluation, and binocular testing at near and at distance.

PRISMS FOR NYSTAGMUS

As early as 1950 Metzger[24] not only insisted on the importance of properly correcting associated refractive errors in congenital nystagmus but also gave the indications for prism prescription: conjugate horizontal prisms for head turn, conjugate oblique prisms for head tilt, base-out prisms for blockage in convergence, and prisms for improving the binocular state of a patient with "occasional manifestation of latent nystagmus." The powers of the prisms prescribed in his three illustrative cases were 10Δ, 6Δ, and 14Δ, in each eye.

A few years later, in 1953, Anderson[2] and Kestenbaum[20] independently published a surgical technique to correct the head turn in nystagmus patients with improved visual acuity in eccentric positions of gaze (neutral zone, null point). In both techniques the eyes are moved in the direction of the head turn either by equal recession and resection in both eyes (Kestenbaum procedure) or by recessions only (Anderson procedure). The technique was soon modified to include the vertical muscles for an abnormal chin position or a head tilt and to take into account the effect of the respective procedure and the muscle operated on. It was reported to be successful, especially for a head turn, in a high percentage of cases* and came to be preferred to the prescription of thick incorporated prisms.

With the goal of preserving one medial rectus for a future procedure, some authors recommended to operate only on three of the horizontal muscles at the first intervention. This induced a secondary strabismus for which a 10Δ to 15Δ base-out prism was needed until it could be surgically corrected.[10;37] Other authors often found the prescription of prisms to be necessary even with

*References 8, 9, 17, 19, 29, 30, 33, 35, 42.

the classical four-muscle Kestenbaum procedure or with Park's modification[29] because of an unplanned induced strabismus or an overcorrected or undercorrected head turn.[15;17] The current trend is to avoid inducing any secondary strabismus and aims, in a single procedure, for an immediate, slight surgical overcorrection of the head turn with final, full correction. With this goal in mind and more prolonged postoperative follow-ups, augmented surgery, of up to 10-mm recessions and resections in both eyes, is now recommended.[5;25;31;32;36] Torsional Kestenbaum procedures[7] are less predictable.[27]

Since compact, reliable, easily operated nystagmographs are now available, nystagmography, by recording and objectifying the movements, has become a very important tool in the diagnosis and management of the different types of nystagmus of concern to the ophthalmologist. Nevertheless, *the response to PAT is invaluable in nystagmus management.* The test not only will reveal whether the abnormal head posture can be corrected or modified by prisms but also can assist in determining the type and amount of surgery to be done to correct the abnormal head posture and any associated strabismus.

Nystagmus without Strabismus and with Improved Visual Acuity in Eccentric Position of Gaze

A patient with nystagmus will turn or tilt the head to favor a gaze direction that dampens nystagmus and improves vision. This results in having the head turned or tilted in the opposite direction to that of the favored gaze and toward the position with the maximum nystagmus and least vision. In the absence of strabismus, conjugate (yoked or homonymous bases), identical power prisms provoke a shift of the gaze without vergence movement and can correct an abnormal head posture while maintaining the best acuity (see Fig. 3-4).

As the image of an object is displaced toward the base of a prism and viewed toward its apex, the bases are placed in the direction of the head turn, tilt, or chin position. Once the favored gaze (neutral zone, null point) is displaced enough by the prisms to reach the center, the patient may straighten the head. For example, with a right head turn, one prism is applied base-in over the left eye, and the other is applied base-out over the right eye. For a chin elevation the two prisms have their bases placed up, and conversely for a chin depression down. For a right head tilt, one prism is placed obliquely (base-out and down) over the right eye, and the other is placed obliquely (base-in and down) over the left eye. Some patients have to be instructed to place their head in the position that allows the best acuity even if somewhat awkward or acrobatic.

If the abnormal position is corrected with prisms, surgery is planned according to the prism power and orientation required to compensate the abnormal head posture. *The eyes are moved surgically in the same direction as the bases of the prisms that is toward the head turn, the chin position, or the tilt.*

In theory, the degree of head turn alone could be used to determine the amount of prism necessary for its correction or the amount of surgery to be done without any prism trial. A perimeter[17(p301)] or other measuring instrument such as an orthopedic goniometer[14;25] or a course protractor compass, as used for navigation, could numerically determine the degrees of the head turn. Because 1° of deviation corresponds to about 2Δ, for dioptric notation the degrees are doubled. However, a head tilt and an abnormal chin position often complicate the turn and render measurements with a perimeter or goniometer impractical. Moreover, with a prism trial often a greater or lesser prism power than calculated by these means may be necessary to fully correct the head turn. This may explain why patients corrected only according to the degrees of the turn can end up with a different result from the same surgery done for the same amount of turn.[1] In most cases the PAT, done in the office for a few hours or in rare instances at home for a few days, will determine more accurately not only if surgery can achieve a full correction but also its type and the amount.

Goddé-Jolly and Larmande reported the interesting case of a patient with nystagmus and improved vision with head turn who, when corrected with prisms allowing best visual acuity with head straight, developed a nystagmus of the head.[17(p407)] It may happen too, when corrected even with a small amount of prism, that the patient will reverse the turn. This patient has a *discordant (disconjugate) nystagmus that was overlooked.* Such a patient is not a candidate for a Kestenbaum[20] type of muscle surgery, unless the fixation was artificially switched and the patient always uses the same eye in casual seeing. There also will be the rare case, in the absence of strabismus, for which paradoxical prisms (bases placed in the direction of least nystagmus) succeed to improve functional vision.[4]

It cannot be emphasized enough that a patient with congenital nystagmus often will benefit from an optical correction even without a substantial improvement in recorded visual acuity. It is essential to do a careful refraction and especially important to fully correct the astigmatism along with the spherical error. Up to 8Δ good quality incorporated prisms, whether used as primary therapy or after surgery, are cosmetically acceptable and may succeed in correcting a head posture when the patient is optically corrected. A stronger prism, which may be necessary to achieve the correction at the beginning of therapy, often can be replaced later by a weaker prism with the same results. In borderline cases, temporary prisms are prescribed to determine the optimal treatment.

Patients who do not benefit from glasses or who wear contact lenses usually prefer to adopt a small head turn rather than to wear prisms. Some even prefer not to assume a compensatory head posture, even if they see less.

Nystagmus blockage in convergence. *In manifest congenital nystagmus convergence is often used as a compensatory mechanism to block the nystagmus.*[11;26;38;40] Exceptionally, it is used in latent and manifest-latent nystagmus.[27;40] For the rare diagnosis of esotropia by blockage of nystagmus in convergence, the angle is variable, the nystagmus intensity is inversely proportional to the angle, and the electrooculographic tracing is the one of *manifest congenital* nystagmus with dampening in convergence. If convergence dampens nystagmus at distance and near, a base-out prism applied over each eye provokes a convergence movement, and when the powers are increased up to the point that the nystagmus decreases, visual acuity improves at distance. The weakest pair of prisms that achieve this goal is prescribed. Base-out prisms, used to elicit convergence, also have been successful to suppress or damp *acquired nystagmus* and to improve vision.[22;43;44] Therefore besides the use of prisms to correct their diplopia, this mechanism of action should be kept in mind to further help these patients. It may be necessary to prescribe minus lenses for best distance vision to nullify the excess accommodation from the prism induced convergence.

Before attempting to create it surgically, artificial divergence should always be induced with prisms.[34;39] With this method, the power of the base-out prisms is increased up to the point where diplopia or exotropia is provoked but does not persist when the prisms are left in place. The amount of prism necessary to elicit this response, which may be as low as 25Δ or up to more than 80Δ, determines the amount of surgery to be done.[41] On the day before surgery, Spielmann has the patient wear Fresnel prism membranes in the power determined, changing them according to the response. She then repeats the PAT at the time of suture adjustment to leave the patient with a convergence reserve of about 20Δ. With this approach, in a series of 120 patients who had medial rectus recession of 5 to 13 mm, consecutive exotropia developed in only nine cases (7%). She found that, although nystagmus blockage in convergence and hypermetropia were only rarely associated, when the latter was present, overcorrections were more frequent.[41]

Associated head turn and torticollis. In some patients a head turn or a torticollis will be corrected when base-out prisms alone are used to induce blockage in convergence.[40(p255);41] Their management is as described above. For the others, who are in the majority, PAT is done to induce convergence and to correct the associated abnormal head posture. For example, for a right head turn, the greater amount of base-out prism is placed over the right eye. If the turn is corrected, a larger recession is done on the right medial rectus than on the left medial rectus. For a blockage in convergence and a chin depression, if base-down prisms added to base-out prisms achieve a correction, then in addition to the medial rectus recessions, recessions of both superior recti are indicated. Some authors recommend incorporated prisms if the powers required are not too high. For example, a patient with dampened nystagmus in convergence and a null point a few degrees left of center had improved visual acuity from 20/40 to 20/25 with 11Δ OD, base to the right, and 3Δ OS, base to the left.[12]

Nystagmus with Strabismus and with Improved Visual Acuity in Eccentric Position of Gaze

The association of strabismus and congenital nystagmus varies according to authors' views and to the type of nystagmus. As stated previously, in manifest congenital nystagmus, the incidence of strabismus parallels that of the general population. However, it is reported to be much higher, up to 85%, with latent and manifest-latent nystagmus. To overlook nystagmus when doing surgery for a strabismus entails serious consequences. The abnormal head posture as well as the nystagmus become more noticeable.[19] Conversely, surgery can be undertaken for nystagmus only after a thorough strabismus evaluation. In particular, the examiner must look for an **A** or **V** pattern for which a chin elevation or depression unrelated to the nystagmus has been adopted. Such patients have an improved binocularity in up or down gaze or a mechanical factor. A torticollis secondary to a paralytic strabismus should also be ruled out, especially when a paradoxical head posture that is not opposite to the null point is present and the prism that corrects the head turn is in reverse of its anticipated position.

In the presence of amblyopia it is important to maintain fixation as in casual seeing during the prism trial test. If this is not done, as fixation is taken with an amblyopic eye, the patient may reverse the head turn, demonstrating a discordant (disconjugate) nystagmus. If in casual seeing the same eye is always used for fixation, the nystagmus is handled as concordant, that is, with the head turn in one direction.

In patients with combined problems the first step is to correct the strabismus, horizontal or vertical, with the appropriate prisms, *taking into account the associated abnormal head posture.* For example, if the patient has a right esotropia with a left head turn *(fixating eye in adduction)*, all the prismatic correction is applied base-out over the left eye, the base of the prism being oriented in the direction of the head turn.

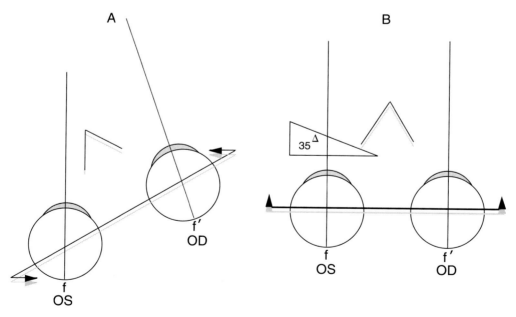

FIGURE 6-1. A, Right esotropia 35Δ with left head turn. Nystagmus dampened with fixating eye in adduction. **B,** Base-out prism over left eye corrects esotropia and head turn. Surgery indicated on left eye.

1. If the prism over the left eye corrects the esotropia and also corrects the left head turn, surgery is done only on that eye (the one with the better vision) to correct the esotropia (Fig. 6-1).
2. If with the esotropia corrected the head turn persists, then a pair of prisms of equal power (conjugate, yoked, homonymous) is introduced, base-out

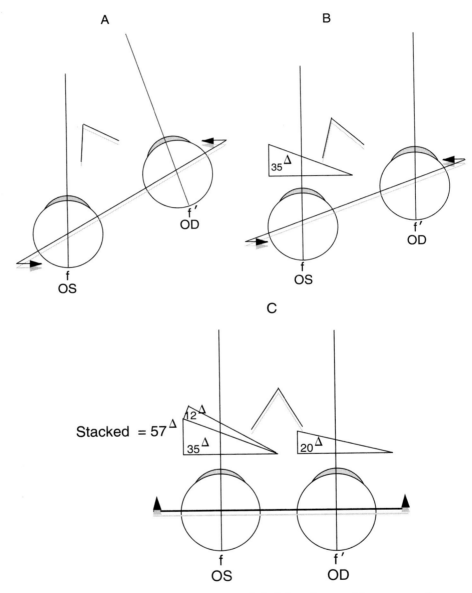

FIGURE 6-2. A, Right esotropia 35Δ with left head turn, Nystagmus dampened with fixating eye in adduction. **B,** Base-out prism over left eye corrects the esotropia but not the head turn. **C,** Conjugate prisms over both eyes, base-out left eye and base-in right eye added to the esotropia correction, correct the head turn. Surgery indicated on both eyes.

over the left eye and base-in over the right eye until the head turn is corrected. If stacked prisms have to be used, their combined power is adjusted according to the stacked prism table. Because the deviation produced by stacking prisms in the same direction is greater than the sum of their labeled values (see Table 1-1) the conjugate prisms will not be of the same labeled values. If prisms approximating 55Δ base-out (stacked 35Δ + 12Δ = 57Δ) are needed over the left eye and a single 20Δ base-in prism over the right eye, surgery is done on both eyes: on the left eye to correct an esotropia of 55Δ and on the right eye to correct an exotropia of 20Δ (Fig. 6-2).

3. With a right esotropia, if the patient dampens the nystagmus with a right head turn *(fixating eye in abduction)*, the esotropia will be corrected by applying the base-out prism over the right eye. If the prism correcting the esotropia also corrects the head turn, surgery on the deviated eye to correct the esotropia will correct the head turn as well (see Fig. 6-3).

4. If, when the esotropia is corrected, the right head turn is still present, then successively, a prism base-in is applied over the left eye and a base-out prism inducing the same power (stacked prism table) over the right eye until both the head turn and the esotropia are corrected. Surgery is planned accordingly.

Some surgeons prefer to correct the strabismus first and manage the head turn in a second procedure. Even so, in the examples chosen, unless an overcorrection is achieved in the first procedure, surgery to correct the head turn in a second procedure will necessarily involve both eyes.

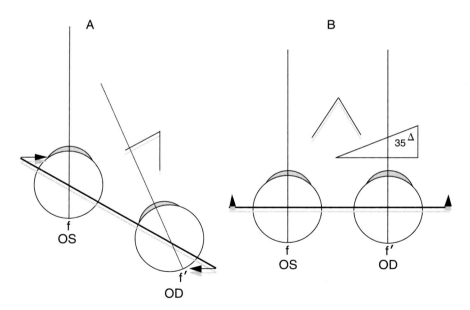

FIGURE 6-3. A, Right esotropia 35Δ with right head turn, Nystagmus dampened with fixating eye in abduction. **B,** Base-out prism over deviating right eye corrects esotropia and the head turn. Surgery indicated on right eye.

From the result of the prism trial test, the examiner will soon discover that when surgery can be performed on one eye only, it will frequently involve the eye with the better vision (nystagmus dampened with the eye in adduction). This should be demonstrated to the parents or the adult patient before surgery by placing over the amblyopic eye the same prism that corrected the turn and the strabismus when applied over the "good" eye. The strabismus will be observed to be still corrected, but the head turn will become worse. The patient or the parents of the child will then be better prepared to accept surgery on the eye with the better visual acuity.

With the fixating eye in abduction, surgery on the deviated eye alone, as in the following case, may suffice. Although such an outcome is infrequent, it illustrates the importance of a prism trial.

- **Case Presentation 6-1**

Patient M.S.: This 15-year-old girl had partially accommodative esotropia with deep amblyopia OS, manifest nystagmus, and left head turn. She had a history of patching of the right eye between 1 and 2 years of age, followed by glasses. Old photos showed a large esotropia with left head turn that was apparent as early as age 3 months. By age 2, and again at age 6, a marked decrease of the deviation was observed.

First exam:
   Visual acuity: (cc) *with left head turn* (nystagmus dampened).
   OD +2.50 +1.00 × 55° = 20/30 near 20/20.
   OS +2.50 +1.00 × 130° = 10/200 near 20/400 (18 cm).
   Nystagmoid fixation OD, eccentric nasal nystagmoid fixation OS.
   Krimsky: (cc) LET 25Δ variable LHT, LET' 25Δ.
   Dextroversion and levoversion within normal.
   RIO⁺, RSO⁻, LIO⁺, LSO⁻.

PAT (1 hour in office):
   Head straight with 20Δ base-out over OS.

Surgery under microscope:
   Resection LLR 12 mm.

Follow-up:
   Same optical correction.
Ten months: patient has maintained head straight with parallelism at
   distance and near.
   Visual acuity: (cc) OD and OU = 20/30, OS = 20/200.

Comments: This patient had a left esotropia and a left head turn. Therefore she was using the abducted right eye for fixation, and a base-out prism could be applied over the deviated eye. Surgery on that eye alone corrected the esotropia and the left head turn as demonstrated during the trial test. Because she needed only 20Δ to achieve this correction, surgery on the left eye was limited to a large resection of the lateral rectus.

REFERENCES

1. Abadi RV, Whittle J: Surgery and compensatory head postures in congenital nystagmus, *Arch Ophthalmol* 110:632-635, 1992.
2. Anderson JR: Causes and treatment of congenital eccentric nystagmus, *Br J Ophthalmol* 37:267-281, 1953.
3. Bietti GB, Bagolini B: Traitement médico-chirurgical du nystagmus, *Année Thér Clin Ophtalmol* 11:268-296, 1960.
4. Bixenman WW: Congenital, hereditary downbeat nystagmus, In Reineke R, editor: *Strabismus II, Proceedings of the Fourth Meeting of the International Strabismological Association, Asilomar, Calif, 1982*, New York, 1984, Grune & Stratton, pp 277-289.
5. Calhoun JH, Harley RD: Surgery for abnormal head position in congenital nystagmus, *Trans Am Ophthalmol Soc* 71:70-87, 1973.
6. Ciancia A: Early esotropia, In Ferrer O, editor, *Ocular motility: Int Ophthalmol Clin*, vol 11, no 4, Boston, 1971, Little, Brown and Co, pp 81-87.
7. Conrad HG, de Decker W: Torsional Kestenbaum procedure: evolution of a surgical concept, In Reinecke R, editor: *Strabismus II, Proceedings of the Fourth Meeting of the International Strabismological Association, Asilomar, Calif, 1982*, Orlando, 1984, Grune & Stratton, pp 301-315.
8. Cooper EL, Sandall GS: Surgical treatment of congenital nystagmus, *Arch Ophthalmol* 81:473-480, 1969.
9. Crone RA: The operative treatment of nystagmus, *Ophthalmologica* 163:15-20, 1971.
10. Cüppers C, Adelstein FE: Posibilidades de tratamiento quirurgico del nistagmus, *Arch Soc Oftal Esp* 32:207-222, 1972.
11. Cüppers C: Historique et physiopathologie des blocages, *J Fr Orthop* 10:15-26, 1978.
12. Dell'Osso LF, Flynn JT, Daroff RB: Hereditary congenital nystagmus: an intrafamilial study, *Arch Ophthalmol* 92:366-374, 1974.
13. Dell'Osso LF: Congenital latent or manifest-latent nystagmus: similarities, differences and relation to strabismus, *Jpn J Ophthalmol* 29:351-368, 1985.
14. Economopoulos NK, Damanakis AG: Modification of the Kestenbaum operation for correction of nystagmic torticollis and improvement of visual acuity with use of convergence, *Ophthalmic Surg* 16:309-314, 1985.
15. Flynn JT, Dell'Osso LF: The effects of congenital nystagmus surgery, *Ophthalmology* 86:1414-1425, 1979.
16. Forssman B: A study of congenital nystagmus, *Acta Otolaryngol* 57:427-449, 1964.
17. Goddé-Jolly D, Larmande A: Nystagmus congénital, In *Les Nystagmus*, vol 1, Paris, 1973, Masson & Cie Editeurs.
18. Helveston E, Ellis FD, Plager DA: Large recession of the horizontal recti for treatment of nystagmus, *Ophthalmology* 98:1302-1305, 1991.
19. Hugonnier R: Notre position actuelle sur les opérations dirigées contre le nystagmus: résultats sur 79 cas, *Ann Inst Barraquer* 12:327-340, 1974-1975.
20. Kestenbaum A: Nouvelle opération du nystagmus, *Bull Soc Ophtalmol Fr* 6:599-602, 1953.
21. Lang J: Special forms of comitant strabismus. In Reineke R, editor: *Strabismus II, Proceedings of the Fourth International Strabismological Association, Asilomar, Calif, 1982*, New York, 1984, Grune & Stratton, pp 770-789.
22. Lavin PJM, Traccis S, Dell'Osso LF, Abel LA, Ellenberger C: Downbeat nystagmus with a pseudocycloid waveform: improvement with base-out prisms, *Ann Neurology* 13:621-624, 1983.
23. Limón de Brown E, Corvera-Bernadelli J: Metodo debilitante para el tratamiento del nistagmus, *Rev Mex Oftalmol*, 63(2):65-77, 1989.
24. Metzger EL: Correction of congenital nystagmus, *Am J Ophthalmol* 33:1796-1797, 1950.

25. Mitchell PR, Wheeler MB, Parks MM: Kestenbaum surgical procedure for torticollis secondary to congenital nystagmus, *J Pediatr Ophthalmol Strabismus* 24:87-93, 1987.

26. Mühlendyck H: Diagnosis of convergent strabismus with nystagmus and its treatment with Cüppers faden operation. In Moore S, Mein J, Stockbridge W, editors: *Orthoptics: past, present, future, Transactions of the Third International Orthoptic Congress, Boston, 1975,* Stratton Intercontinental Medical Book Corp, New York, 1976, pp 143-154.

27. von Noorden GK: *Binocular vision and ocular motility,* ed 4, St Louis, 1990, Mosby, pp 435-454.

28. von Noorden GK, Sprunger DT: Large rectus muscle recessions for the treatment of congenital nystagmus, *Arch Ophthalmol* 109:221-224, 1991. In reply to letter to editor related to above paper: *Arch Ophthalmol* 109:1636-1638, 1991

29. Parks MM: Congenital nystagmus surgery, *Am Orthop J* 23:35-39, 1973

30. Pierse D: Operation on the vertical muscles in cases of nystagmus, *Br J Ophthalmol* 43:230-233, 1959.

31. Pratt-Johnson JA: The surgery of congenital nystagmus, *Can J Ophthalmol* 6:268-277, 1971.

32. Pratt-Johnson JA: Results of surgery to modify the null-zone position in congenital nystagmus, *Can J Ophthalmol* 26:219-223, 1991.

33. Raab-Sternberg A: Anderson-Kestenbaum operation for asymmetrical gaze nystagmus, *Br J Ophthalmol* 47:339-345, 1963.

34. Roggenkamper P: Combination of artificial divergence with Kestenbaum operation in cases of torticollis caused by nystagmus, In Reineke R, editor: *Strabismus II, Proceedings of the Fourth International Strabismolological Association, Asilomar, Calif, 1982,* New York, 1984, Grune & Stratton, p 329-334.

35. Schlossman A: Nystagmus with strabismus: surgical management, *Trans Am Acad Ophthalmol Otolaryngol* 76:1479-1486, 1972.

36. Scott WE, Kraft SP: Surgical treatment of compensatory head position in congenital nystagmus, *J Pediatr Ophthalmol Strabismus* 21:85-95, 1984.

37. Sevrin G, deCorte H: L'usage des prismes dans le nystagmus, *Ann Ocul* 203:437-443, 1970.

38. Spielmann A: Indication du traitement chirugical dans les strabismes avec nystagmus, In Bérard PV, Quéré MA, Roth A, Spielmann A, editors: *Chirurgie des Strabismes,* Paris, 1984, Soc Franc Ophtalmol et Masson, pp 413-463.

39. Spielmann A, Dahan A: Double torticollis and surgical artificial divergence in nystagmus, In Nemet P, Weiss JB, *Acta Strabologica, Proceedings of the International Symposium on Strabismus and Amblyopia, Tel-Aviv 1985,* Paris, 1985, CERES, pp 187-192.

40. Spielmann A: *Les strabismes—de l'analyse clinique à la synthèse Chirurgicale,* ed 2, Paris, 1991, Masson, pp 234-285.

41. Spielmann A, Laulan A: La mise en divergence artificielle dans les nystagmus congénitaux. A propos de 120 cas, *Bull Soc Fr Ophtalmol* 93(6/7):571-578, 1993.

42. Taylor JN: Surgery for horizontal nystagmus, the Anderson-Kestenbaum procedure, *Aus J Ophthalmol* 1:114-116, 1973.

43. Traccis S, Rosati G, Monaco MF, Aiello I, Pirastru MI, Becciu S, Loffredo P, Agnetti V: Alternating esotropia, monocular and binocular square wave jerks. Improvement with base-out prisms, *Neuro-Ophthalmol* 8:43-49, 1988.

44. Traccis S, Rosati G, Monaco MF, Aiello I, Agnetti V: Successful treatment of acquired pendular elliptical nystagmus in multiple sclerosis with isoniazid and base-out prisms, *Neurology* 40:492-494, 1990.

45. *Walsh and Hoyt' clinical neuro-ophthalmology,* ed 4, Miller NR, Baltimore, 1985, Williams & Wilkins.

CHAPTER 7

Incomitant Deviations

The main goals of prism prescription in incomitant deviations are to relieve the patient of diplopia, to correct a significant abnormal head posture, and to allow single vision *in primary position and down gazes.*

General approach:

1. Determine the minimum amount of prisms to achieve the above.
2. Take into account that adults in general will only tolerate a Fresnel prism membrane over the nonpreferred eye.
3. Use oblique prisms for combined deviations.
4. Be prepared to use different powers of prisms in the same patient over the same eye.
5. Always conduct an office trial to determine tolerance.
6. Personally apply or confirm the accurate application of the membrane. Often some rotation and adjustments are needed.
7. In the presence of a torsional residual diplopia, give a home trial. Fusion may compensate for it with time.
8. Recommend that the patient adjust to the prismatic correction in a secure environment first.

A field of binocular fixation (diplopia field) is useful to document the results achieved with the prismatic correction.

Congenital Incomitant Deviations

In the child, prisms are rarely prescribed for congenital incomitant deviations, whether induced by a true paralysis or by structural anomalies, unless an optical correction is also indicated and the compensatory head posture can be corrected with a prism of low power. Conjugate base-up prisms may help a child with a Brown's superior oblique syndrome to compensate a chin elevation when the parents refuse surgery. A small vertical prism may correct a face turn as well.[4] When prisms are prescribed for a retraction syndrome (Stilling, Turk, Sinclair, Duane) the parents have to be informed that the deviation and the face turn may increase spontaneously with time, necessitating corrective surgery despite initially successful prism correction. Before surgery a prism test is done with the base of the prism in the direction of the face turn. Its power is increased to achieve a correction of the deviation as well as the

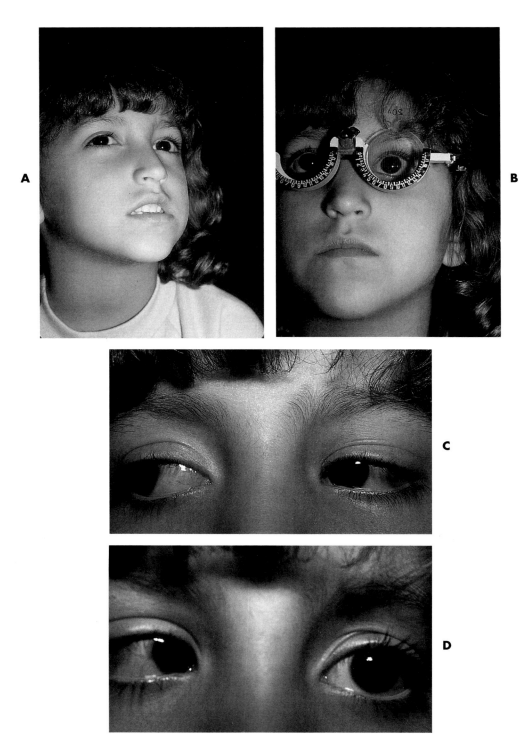

FIGURE 7-1. **A,** Duane's retraction syndrome of the left eye with left face turn. **B,** Esodeviation and left face turn corrected with 20Δ base-out Fresnel trial set prism over left eye. **C,** Dextroversion. **D,** Levoversion.

face turn. This test and the Hess-Lancaster graph are the best ways to demonstrate to the patient or the parents the correction that can be achieved surgically in primary position while the inherent muscle limitations are still present (Figs. 7-1 and 7-2, A and B).

In the adult an old decompensated ocular torticollis is often mistaken for a recently acquired paralysis. The patient suddenly starts to see double because of a change in visual or postural habits and often denies a preexisting strabismus or torticollis. The sensory state, the history, and above all the study of a series of old photographs from earlier years facilitate the diagnosis.

Patients with old decompensated torticollis have large vertical amplitudes of fusion that have become insufficient to maintain binocularity but are not easily helped with prisms because of secondary musculoskeletal changes. They not only tend to return to the abnormal head posture but also can find a prismatic correction useless or at best disturbing. On the other hand, some patients will respond well to a gradual increase of the vertical correction, become asymptomatic when partially corrected, or be helped with prisms only in their reading glasses.

It should be kept in mind that an old paralysis may mask an evolving lesion, especially if a progression in the compensatory head posture is documented. When in doubt a neurologic evaluation is in order.

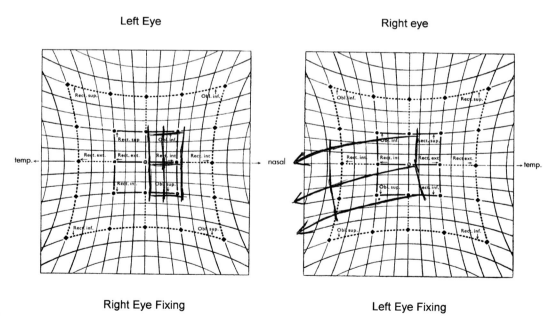

FIGURE 7-2. A, Hess-Lancaster graph with head straight demonstrates limitation of abduction of the left eye as well as adduction not apparent in dextroversion (see 7-1, C).

Left Eye

Right eye

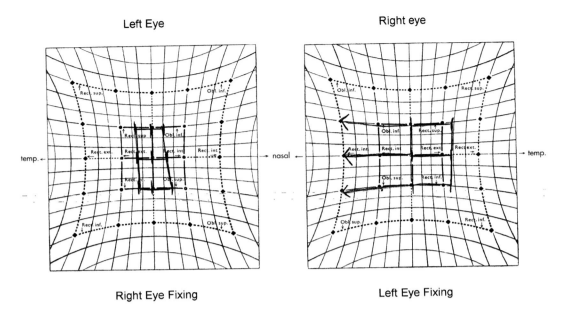

Right Eye Fixing

Left Eye Fixing

FIGURE 7-2. B, Hess-Lancaster graph. With prism or after a 7 mm recession of the left medial rectus similar results: face turn corrected, no deviation in primary position, but as predicted, limitation of adduction and abduction persists in the left eye.

Acquired Incomitant Deviations

Recent acquired palsy (paralysis, paresis). Diplopia is almost always present in extraocular muscle paralysis of recent onset. Whether the etiology can be determined or not, it is good medical practice to wait a minimum of 6 to 8 months before considering a surgical correction. Therefore the diplopia will need to be managed for that period at least. In acute sixth nerve palsy, botulinum toxin[35] (Botox) injections in the ipsilateral antagonist (medial rectus) during the observation period will correct diplopia totally or partially. The effect of the injection lasts for about 3 months. Even though Huber[22] has reported good results in fourth nerve palsy and substantial improvement with third nerve palsy, botulinum toxin injections frequently fail to fully compensate these patients. With sixth nerve palsy some will refuse this form of treatment. Although Fresnel prism membranes have been available for many years, it is surprising that all too often a patient is given only the choice of wearing an occluder during the observation period. Frequently the treating physician has been led to believe that the diplopia cannot be satisfactorily compensated with prisms in incomitant deviations or that they are useful only if the deviation is less than 10Δ[8;33] or involves only the horizontal muscles.

However, it has been my experience[37] with others that most cases of recent paralysis, even large deviations up to 40Δ and combined deviations, can be successfully neutralized and diplopia relieved with Fresnel prisms. First, the examiner must ensure that the diplopia is binocular and not monocular.

Then the nature of the diplopia—horizontal, vertical, or torsional—is determined. Is it present at near and distance, and in what position of gaze is it most annoying? The examiner should observe which eye the patient spontaneously closes to avoid diplopia, because it is habitually the affected eye over which the prism will be applied.

Office prism trial. The examiner starts with the position where the diplopia is most annoying to the patient. The patient is informed that a "successful prism" will be the one that can overcome his or her diplopia in straight-ahead gaze at distance, as well as at near in the reading position, and that sometimes two different prisms may have to be used to achieve this effect. *The patient must understand that diplopia will not be corrected in all directions of gaze.*

The least amount of prism that overcomes the diplopia in the straight-ahead position or in the reading position or both is determined. If there is an associated horizontal and vertical deviation, the use of an oblique prism will allow the correction of both with a single membrane. To determine the oblique prism requirement a prism of a power greater than the larger deviation but less than the two deviations combined is chosen. With the apex of the prism placed toward the direction of the deviation, the prism is rotated to the point where both the horizontal and the vertical diplopia are overcome. This subjective method for determining the axis position of an oblique prism is the most practical in this group of patients (see Chapter 1). Once the power and axis are determined, a round loose prism, or for powers above 12Δ, a prism from the Fresnel trial set, is taped over one spectacle lens or fixed in a trial frame over the paralytic eye or nonpreferred eye for the office trial.

In my experience, an adult patient rarely tolerates a Fresnel membrane-over both eyes or over the preferred eye. Any intolerance to a Fresnel prism from the trial set is especially meaningful because a Fresnel membrane will induce even more distortion and vision loss.

Prism prescription. To correct diplopia at distance, straight ahead, and reading positions, once the power and orientation of the prism are determined by the office trial, the membrane is applied to the patient's spectacles or, in the absence of the latter, to plano sunglasses (see Technique of cutting and applying the Fresnel prism membrane described in Chapter 2). If a deviation is significantly different at distance and near, two prisms of different powers and orientation can be applied to the lower and upper segments of the same spectacle lens. Even patients with deviations greater than 30Δ may still overcome their diplopia with a single 30Δ membrane using either fusion or a small abnormal head posture to compensate for the residual deviation. Many prefer the prism correction to an occluder that restricts their field of vision. A membrane can also be added to glasses that already have incorporated prisms. The combined power may not correspond exactly to the true angle of deviation, but this is of little importance to the patient unless the combined prism powers are used to determine the angle for surgery (see Chapter 1). The combination of spectacle-incorporated prisms, divided equally between both eyes, and a Fresnel membrane over the nonpreferred eye also allows for the use of a lower power membrane in patients who cannot tolerate or adjust to the distortions or chromatic aberrations induced by a higher power membrane. Membranes of 35Δ and 40Δ, now available, are rarely tolerated.

Because of the reduction in visual acuity it induces, it may be argued that a prism membrane of 30Δ or of greater power acts as an occluder to overcome diplopia rather than by reestablishing fusion. Binocularity can be verified by decreasing or increasing the prismatic correction with a prism bar passed in front of the eye without the membrane. If diplopia is provoked, the Fresnel membrane is not acting as an occluder. Moreover, poor vision in one eye does not necessarily protect against diplopia.

Effect of prisms on evolution of the palsy. The adult patient with a monocular palsy and diplopia will prefer the Fresnel prism over the paralytic eye because the deviation is smaller (primary deviation) and past-pointing can be overcome more readily when fixating with the nonparetic eye. For this reason, the patient will often spontaneously close the paretic eye before treatment is instituted. The exception to this rule occurs when the vision in the paretic eye is significantly better than that in the nonparetic eye.

The application of prisms over the nonparetic eye has been advocated to prevent the development of contractures in the paralytic eye.[16] However, no studies have proven that wearing prisms over either the paretic or the nonparetic eye hastens the recovery of the paralysis. A large percentage of such patients recover spontaneously and often do so in a very capricious way. It is nearly impossible to objectively evaluate any therapeutic effect from the wearing of prisms other than the mechanical improvement of the position and size of the field of binocular fixation (Case Presentations 7-1 and 7-2). Prisms are used primarily as a crutch to improve function and are usually preferred by the patient to an occluder which eliminates diplopia but at the expense of binocular vision.

- **Case presentation 7-1**

  Patient P.M.: This 65-year-old man with diabetes had a left sixth nerve palsy. He reported horizontal diplopia at distance in all fields of gaze, except in right gaze, since 3½ months. He had been told that the diplopia would disappear. To function at distance he either closed one eye or chose the "right (OK)" image of the two with which he was able to maintain orientation. At near he was able to eliminate the diplopia by adopting a slight left face turn.

  First exam:
  Visual acuity (pc):   OD +3.00 −0.50 × 108° = 20/25.
  OS +3.50 +0.75 × 80° = 20/25.
  Add +3.00 OU = J1.

  ACPT (OD fixing): LET 20, E' 6Δ. Incomitant.
  Right gaze: E 2Δ, Left gaze: LET 40Δ.
  Ductions: marked limitation in abduction OS.

Prism prescription:

The patient had no diplopia with a 20Δ base-out Fresnel trial prism but was more comfortable with 30Δ base-out. A membrane of this power was applied to the upper segment of his left bifocal lens.

Follow-up:

Two months: (5½ months after onset of palsy).

Occasional recent diplopia with glasses. Slight right face turn.

ccΔ: LXT 30Δ.

cc no Δ: LET 6Δ, X' 4Δ.

Right gaze: ⊕.   Left gaze: LET 12Δ.

5Δ base-out, necessary to overcome diplopia in primary position at distance, replaced 30Δ over the upper segment of left bifocal lens.

Five months: asymptomatic. Functions better with glasses to which Fresnel has been applied, compared to pair without prism.

cc no Δ: LET 4Δ, X' 4Δ.

Right gaze: ⊕.   Left gaze: LET 6Δ.

Stereoacuity: 40 sec of arc at near (Titmus).

60 sec of arc at distance (Vectograph).

Incorporated prisms prescribed 2Δ base-out each lens.

Nine months: has continued to wear old Fresnel prism membrane.

Motility unchanged. The membrane was removed a few months later.

Three Years:

ACPT (cc): orthophoria in primary position as well as in right and left gaze.

At near X' 6Δ.

Comments: This patient, who was without any prismatic correction for 3½ months after onset of his sixth nerve palsy, improved suddenly after wearing a strong prism for only 2 months over the upper segment of his bifocal lens on the paretic side. He fully recovered from his palsy about 9 months later and meanwhile needed only a weak prism to be comfortable. Although the prismatic correction may have had little to do with the evolution of his palsy, it allowed the patient to function binocularly and comfortably—advantages he could have enjoyed from the onset of his palsy if a prism had been prescribed.

- **Case Presentation 7-2**

Patient I.S.: This 62-year-old man reported the sudden onset of horizontal diplopia at distance for the past few days. He occluded his right eye, and he was under numerous circulatory medications. A neurologic work-up showed normal results.

First exam:

Visual acuity (pc): OD −9.00 = 20/40 −6.50 = J2.

OS −12.00 = 20/40 −9.50 = J2.

Incomitant RET. Ductions showed a marked limitation of abduction of the right eye.

ACPT (cc): (FOD) ET 40Δ, (FOS) ET 30Δ.

Prism perscription:

After office trial with a 30 base-out Fresnel trial set prism which overcame his diplopia at distance and near a Fresnel membrane of the same power and direction was applied to the right lens of both distance and near glasses.

Follow-up:

One month: patient preferred prism to occluder.

No diplopia.

Two months: condition unchanged

Three and one-half months: with Fresnel membrane visual acuity
OD reduced from 20/40 to 20/100.

ACPT (cc no Δ): FOD 16Δ, FOS 14Δ, E' 4Δ.

Prism reduced to 10Δ base-out on the right lens and removed from reading glasses.

Five months: same prescription.

Nine months: *prism removed.* No diplopia.

Maddox rod: (FOS) E 10Δ RH 2Δ, X' 4Δ ⊖'.

Three years: (cc) X 2Δ, X' 14Δ.

Comments: The office prism trial for this high myopic patient revealed that the same prismatic correction at near and distance corrected his diplopia. A 30Δ Fresnel membrane was placed over his paretic eye only a few days after the onset of his sixth nerve palsy. Although it is probable that the early prismatic correction did not shorten his recovery from the palsy, the immediate correction of the diplopia allowed him to function comfortably from its onset to its resolution.

Prisms Used for Orthoptics in Paralytic Strabismus

For more than a century prismotherapy has been advocated for the management of paralytic strabismus. In 1864, in the *Journal of London Hospital Medicine and Surgery*, Ernest Hart[21] published a "Case of extreme squint cured without operation by the use of prisms." The patient had a right sixth nerve palsy. In primary position the diplopia could be overcome with a 12° glass prism. With a prism of 10° the images could be nearly fused but not quite. To gymnastically exercise and strengthen the enfeebled muscle, the 10° glass prism was selected for the patient to wear *over the deviating eye.* The patient found the effort to maintain single vision painful and could not sustain it for long. Therefore Hart directed the patient to use the prism at intervals, gradually lengthening the period of exercise. In the course of 7 days, the patient had made considerable progress. Exercises were continued with prisms of lesser degrees. In 8 weeks the patient was "cured." This method was advocated for paralytic strabismus, where the amount needed to overcome the diplopia was less than 12°, and was not advocated for patients who could overcome the diplopia spontaneously.

Over the years there have been periods of enthusiasm and neglect for vergence exercises involving prisms or other instruments. The unpredictable evolution of an ocular palsy, which often can remain stationary for some months and then disappear in a few days without treatment, has left few advocates of vergence exercises for their management.

Guibor's technique. Guibor, who revived prismotherapy for strabismus in the 1940s, had an elaborate technique combining occlusion and prisms.[15;16] His technique, described especially for a sixth nerve palsy, is mentioned here because it is often confused with the simple application of a prism over the nonparalytic eye. Guibor's technique comprised the following steps. (1) During the first week the nonparalytic eye was covered, and the total prismatic correction, less 10Δ, was applied to the paralytic eye. (2) During the second week, constant occlusion of the nonparalytic eye was replaced by intermittent occlusion. The power of the prism over the paralytic eye was decreased gradually while a prism over the nonparalytic eye was applied and its power increased proportionally.

The goals of his technique were "limitation of suppression" and the prevention of contracture, as well as reduction of the diplopia and decreased head rotation.

I do not follow Guibor's technique for the following reasons[37]:
1. The patient tolerates poorly the occlusion of the nonparalytic eye; the deviation becomes greater, and the symptoms become worse.
2. Often the nonparalytic eye is the better eye visually. Because the Fresnel membrane decreases vision, it is better applied to the nonpreferred eye.
3. In the adult, suppression is of little concern. On the contrary, it is unfortunate that in intractable cases of diplopia, the adult does not suppress more easily.
4. The overall technique is difficult to follow, and its effect is even more difficult to verify because, in sixth nerve palsy, recovery is often spontaneous.

Cyclodeviations (Torsional Deviations)

Torsional diplopia is a frequent symptom in four nerve palsy of recent onset. It can also accompany paretic and mechanical limitation of all vertical muscles. Although a cyclotropia of up to about 15° (Maddox rod or Synoptophore measurements) can be found in long-standing deviations, the patient rarely complains of it. Through a central mechanism, good fusional amplitude, and head position, this patient succeeds in compensating for it with time.

Cyclotropia cannot be corrected with prisms. However, when the vertical and horizontal deviations can be neutralized by a prismatic correction, if the patient still complains of a residual torsional diplopia and surgery is not an immediate option, a trial with a Fresnel membrane should be given. With time the torsional diplopia may be overcome, and the symptoms will disappear (Case Presentation 7-3).

- **Case Presentation 7-3**

  Patient R.J.: This 70-year-old woman had horizontal and vertical diplopia secondary to thyroid ophthalmopathy. She was told that prisms would not help. This patient was unhappy with the occluder that she was using full time restricting her field of vision as well as interrupting her binocularity.

  First exam:
  Visual acuity (pc): OD $-1.25 \, -0.50 \times 90° = 20/30$.
  OS $-1.50 \, -1.00 \times 90° = 20/30$.
  Limitation of elevation OD (dominant eye).
  Maddox rod: ET 6Δ LHT 18Δ excyclo 15°,
  E′ 1Δ LHT′ 25Δ excyclo 15°.

  Prism prescription:
  A 20Δ Fresnel trial prism apex at 120° OS overcame her diplopia at distance. A membrane of the same power and direction was applied on the left lens of her distance prescription.

  Follow- up:
  Two weeks: the patient reported that she was very happy with the oblique prism because of great improvement in her field of vision and binocularity, especially outdoors, and no need for an occluder. Diplopia was found in up gaze only. The same prism Rx was applied to her tinted glasses (Fig. 7-3). After office trial a 15Δ base-down Fresnel prism was applied to the left lens of her reading glasses.

  Two years and eight months:
  No complaint of diplopia with prism corrections at distance and near.
  Horizontal, vertical, and cyclodeviations unchanged.
  Surgery was delayed because of her poor state of general health. Prism correction was well accepted by the patient.

  --------------------- Δ ----------------------

Diplopia Secondary to Central Disruption of Fusion

A patient with diplopia secondary to central disruption of fusion, a rare pathology, usually has a history of severe closed head trauma with coma and intractable diplopia. This pathology has been described as a sensory fusion disruption syndrome.[25] For von Noorden, the pathology is an acquired motor fusion deficiency.[33(p433)] He has observed the marked decrease or complete absence of fusional amplitude but the presence of fusion and stereopsis during the brief moment that the patients were able to superimpose the double images. The most common observation is that of a patient who was previously able to fuse and who is now unable to superimpose, and if so only briefly and without fusion when the diplopia images are precisely aligned with prisms or with a haploscopic device like the major amblyoscope. Bilateral superior

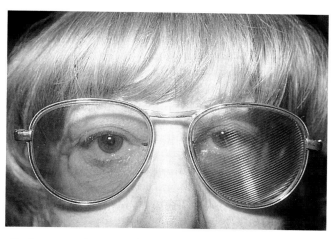

FIGURE 7-3. Patient wearing over her nonpreferred left eye a 20Δ Fresnel membrane apex at 120° for an esotropia of 6Δ and a left hypertropia of 18Δ. She was able to compensate for an excylclotropia of 15° after wearing this oblique prism for 2 weeks. Note that the patient's left eye is seen to be displaced up and nasally in the direction of the apex of the correcting prism.

oblique palsy may also be acquired from closed head trauma.[24;27] However, unless associated with central disruption of fusion, it will allow fusion to be reestablished when the excyclotropias are corrected. *Therefore it is important to correct the full deviation, including any associated cyclotropia, before making the diagnosis of central fusion disruption.* It is equally important to evaluate the potential for fusion in different directions of gaze and to look for a field defect that could account for the absence of fusion. The patient has neither a subjective angle of fusion nor a zone of suppression. Therefore the condition can be easily differentiated from diplopia associated with a change of angle in an old strabismus. A complete history and general eye examination will eliminate other causes of acquired intractable diplopia related to eye conditions such as a preretinal membrane or aniseikonia.

Although reporting on diplopia that resulted from a different etiology, Pratt-Johnson and Tillson have included in this entity patients with unilateral traumatic cataract acquired after age 10, corrected surgically or optically after an interval of at least 3½ years, who had intractable diplopia and no fusion.[34]

The prognosis for recovery in head trauma based central disruption of fusion is poor. Pratt-Johnson and Tillson have reported partial recovery in three patients many years after the accident. London and Scott[25] reported the recovery of fusion in a 17-year-old patient after 4 months of wearing Fresnel prisms for an esotropia of 20Δ and a hypertropia 4Δ, which initially had placed the unfused images over each other.

Restoration of parallelism of the visual axes does not seem to have been a precondition for recovery in the few successful cases reported. In any event the clinician has little to offer besides occlusion. In the highly motivated patient a prism trial to neutralize the objective angle is worth a try.

Myasthenia Gravis

Myasthenia gravis, a neuromuscular disorder characterized by weakness and fatigue of the skeletal muscles, is not rare, having a prevalence rate of at least 1 in 10,000.[20] The underlying defect is a decrease in the number of available acetylcholine receptors at the neuromuscular junction, resulting from an antibody-mediated autoimmune attack.[20]

The lids and extraocular muscles, involved initially in about 70% of patients, eventually become affected in over 90% of patients. Hence the ophthalmologist may be the first physician consulted and often makes the diagnosis. The disease can mimic a partial or total third, fourth, and sixth nerve palsy, although torsional diplopia is a rare sign. The disease can also simulate an internuclear ophthalmoplegia. It occurs in women twice as often as in men, and although more common in the second decade of life, the disease may strike at any age in both sexes. The prevalence of associated thyroid disorders (hypothyroid and hyperthyroid) may be as high as 9% in men and 18% in women. The linkage between the two diseases is probably genetic and immunologic.[41]

Diplopia, lid twitch, and ptosis are common initial complaints of myasthenia. Difficulty accommodating is less frequent. The diagnosis is confirmed by a resolution of the ocular findings within 3 minutes of an intravenous injection of edrophonium (Tensilon) or within 45 minutes of an intramuscular injection of neostigmine methylsulfate (Prostigmin). The latter test is more suitable in children and in addition allows a comparison of the measurements with the alternate cover and prism test in the diagnostic positions of gaze or a Hess-Lancaster graph before and after the test. Even if these tests are negative, the diagnosis can be ruled out only after results of electrophysiologic, serologic, and muscle biopsy with receptor assay testing are found to be negative.

Although the current treatment is said to be highly effective,[20] persistent ocular symptoms have been reported in 40% of patients with an average follow-up time of 14 years (range is 1 to 39 years).[2] These individuals should always be offered prisms as an alternative to an occluder. Some patients report a variable diplopia that may be absent or minimal on awakening, but becoming worse as the day progresses and as muscular activity increases. In these patients, two or three pairs of glasses with different prismatic corrections, prescribed according to the severity of the diplopia, may be of benefit.[30] Others with constant angle deviations are managed with a single prismatic correction. Conjugate vertical base-up prisms have also been found useful to correct a compensatory chin elevation in these patients.[9] Initial difficulty with accommodation usually resolves with the medical treatment. However the asthenopia may still be helped and poor convergence improved with base-in prisms prescribed for near vision (see Chapter 5). As a rule, strabismus surgery is not indicated unless the ocular deviation has been stable for a long time in a patient with the myasthenic condition otherwise under control. For this select group of patients, prisms may still have to be used when there is a relapse of asthenopic symptoms or diplopia.

Thyroid Ophthalmopathy

Graves' disease, also known as Parry's or Basedow's disease, is a disorder with a triad of major manifestations: hyperthyroidism with diffuse goiter, ophthalmopathy, and dermopathy. The three tend to run courses that are largely independent of one another both in time and severity. It is a relatively common disease, affecting 1% to 2% of the general population, that occurs at any age but especially in the third and fourth decade. The predominance in women compared with men may be as high as 7:1.

Graves' disease is an autoimmune thyroid disease of unknown cause.[19] One proposed mechanism of the ophthalmic component is the development of antibodies against specific antigens in the extraocular muscles. Ophthalmic manifestations, that can be seen in the hyperthyroid, euthyroid, and even hypothyroid patients (after treatment), consist of edema of the orbital tissues with enlargement and progressive fibrosis of the extraocular muscles, proptosis, lid retraction, exposure keratitis, secondary increase of intraocular pressure, and neuropathy by compression.

The inferior rectus is the eye muscle most often involved, but with time the pathology can affect all the ocular muscles to a variable degree. A magnetic resonance imaging (MRI) evaluation is useful in objectively determining the extent of the muscle involvement. The deviation can reverse, although this reversal is more prone to happen after muscle surgery.[14;23] The reversal can be due to involvement of a yoke muscle or the improvement of the originally affected muscle. The fibrotic process that follows the myopathy is not irreversible, but significant regression of a large deviation is infrequent. Therefore the patient should be given a prism, rather than an occluder, and the alternative of surgery only later when the condition has stabilized.

Patients with dysthyroid eye disease may have an abnormally large fusional range, particularly in the vertical plane.[14] Thus the amount of prism required to correct their diplopia can be less than the objective angle would indicate. Contrary to true paralytic strabismus, because of the mechanical factors present, patients may prefer to take fixation with the affected eye, for example, with the hypotropic eye. Again, this points to the necessity of performing a prism trial before applying or prescribing prisms.

It is not infrequent, with a thyroid-induced fibrosis of the inferior rectus, to find a patient who has no diplopia when reading but has intractable diplopia in straight-ahead gaze. In this instance the correcting Fresnel membrane should be cut to cover only the upper segment of the glasses. Disparate power membranes are also applied to the bifocals in patients who have a different manifestation of their strabismus at near and at distance (see Fig. 7-4 and Case Presentations 7-4 and 7-5).

When the patient is well controlled medically, has a prism-compensated deviation, is stable for at least 6 months, and desires to be relieved from wearing the high-power prisms, surgery can be planned. Although the patient had to wear an oblique prism before surgery, in the dysthyroid patient with a large esotropia and a small vertical deviation, in my opinion it is better to correct the esotropia surgically and to leave the vertical uncorrected. Often no prism

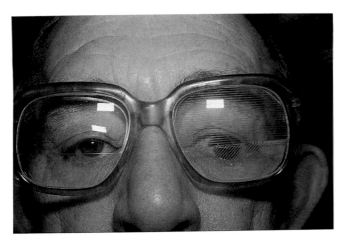

FIGURE 7-4. Patient F.G. wearing on his bifocal lens Fresnel prisms of disparate power and orientation over the left eye with less vision. On upper segment, 20Δ base-up membrane, and on lower segment, 30Δ membrane at 145° to correct an exotropia of 35Δ and a hypotropia of 14Δ. With this arrangment, he had no diplopia at near and at distance. Note the displacement of the left eye in the lower segment in the direction of the prism apex.

or only a small prism will be required postoperatively to maintain fusion, and a vertical overcorrection will be avoided. The same recommendation applies when surgically correcting large vertical deviations with associated horizontal deviations. Forced ductions, done at the time of surgery, do not always determine the overall effect that the release of the greater mechanical restriction will have on associated lesser restrictions. Even with adjustable suture technique, postoperative prisms will often be necessary to achieve alignment and a good functional result.[26]

Different high-power membranes in upper and lower segment of one bifocal lens

- **Case Presentation 7-4**

Patient F.G.: This 59-year-old man had severe thyroid ophthalmopathy and constant horizontal and vertical diplopia. Two years earlier he had bilateral orbital decompression, followed 6 months later by bimedial recessions and right inferior rectus recession. A Fresnel prism eliminated the vertical diplopia at distance only. He was told his near diplopia was uncorrectable because of torsion, and he wore a patch when reading.

First exam:
    Visual acuity (pc): OD $-0.50 -2.00 \times 150° = 20/20$.
                          OS $-0.50 -1.50 \times 180°$ with 15Δ base-up Fresnel prism above reading segment = 20/40. No Δ = $20/25^{-2}$. Add +2.25 OU = J1.
    ACPT (cc no Δ): RHT 16Δ. Incomitant.
    Through near segment: XT′ 35Δ RHT′ 14Δ, excyclo 4° OD.

Office prism trial:

At distance: patient was more comfortable with 20Δ base-up Fresnel trial
   prism over OS.

At near: horizontal, vertical and torsional diplopia corrected with 30Δ
   trial prism base at 145° over OS.

Prism prescription:

20Δ base-up Fresnel membrane applied to upper segment of left lens.

30Δ base at 145° applied to lower segment of left lens. (see Fig. 7-4)

Comments: For 1 year, up to the time of corrective surgery, which had to
be delayed for medical reasons, he experienced no diplopia with prism
glasses and had fusion at distance and near with the Worth 4-dot test.

   Because of his quick adaptation to the new prism prescription, fitted at
his first visit to our office, the patient questioned as to why he had not
received a similar "successful prismatic correction" when he consulted at
the onset of his diplopia or at one of his many earlier follow-up visits
"elsewhere."

Incorporated high-power prisms versus a Fresnel membrane

• **Case Presentation 7-5**

Patient L.L.: This 60-year-old man had a horizontal diplopia that had
started 6 months previously. It had been intermittent at onset and was more
evident in down gaze than in right gaze. It was now constant, could not be
overcome with posture, and was very annoying. He had an office prism trial
elsewhere but "the examiner gave up!" He had a history of thyroidectomy
for Grave's disease 27 years earlier and was euthyroid. Ultrasonograms
suggested a thyroid myopathy.

First exam:

   Visual scuity (sc):   OD 20/25 +0.50 = 20/20.
                         OS 20/25 +0.50 = 20/20.
                            Add +2.50 OU = J1.

   Limitation of abduction OD.

   No head posture could compensate diplopia.

   ACPT: ET 18Δ   LHT 3Δ,   ET' 20 LH' 2Δ.

   Decreased to ET 8Δ and ET' 12Δ in left gaze, about the same in other
   gazes. No cyclodeviation.

Office prism trial:

   At distance diplopia was overcome with 25Δ Fresnel trial set prism, base
   165° OD.

   At near patient was comfortable with a total of 12Δ base-out.

Prism prescription:

   25Δ Fresnel prism membrane base 165° applied to right lens of distance
   spectacles. 6Δ base-out prism incorporated in each lens of reading glasses.

Follow-up:

Three months: no diplopia with either correction, but patient annoyed by distortion and reduction in vision induced by the Fresnel membrane: OD 20/40 to 20/50 with left face turn.

Fresnel membrane replaced with high power incorporated prisms:

OD +0.50   11 base-out.

OS +0.50   11 base-out.

Overcame small vertical component.

Follow-up:

Six months: happy with incorporated horizontal prisms. Can drive his car and do his work as an accountant with comfort. Does not want surgery.

Maddox rod (ccΔ):   ① LH 4Δ, ①′ LH′ 1Δ

Patient's condition was followed by his local ophthalmologist.

Six years: patient has developed glaucoma OS.

Applanation tonometry readings were not affected by eye positions.

Little change in horizontal deviation, but the vertical deviation has reversed. The patient now has an RHT 12Δ and RHT′ 10Δ with a small excyclophoria.

Fresnel membrane combined with high-power incorporated prisms.

Distance prescription: (Fig. 7-5 A).

OD +0.50 12Δ base-out (inc).

OS +0.50 13Δ base-out (inc) and 12Δ base-up (membrane).

Near Prescription: (Fig. 7-5 B).

OD +2.50 6Δ base-out (inc) and 12Δ base-down (membrane).

OS +2.50 6Δ base-out (inc).

Follow-up:

Two months: no diplopia, ccΔ:

Vectograph: suppresses OS.

Titmus: 3000 sec of arc.

Maddox rod: RH 3Δ   X 1Δ,   RH′ 3Δ   X′ 1Δ.

Surgery accepted.

PAT (in office):

ET 30Δ   RHT 20Δ,   ET′ 26Δ   RHT′ 20Δ.

Surgery cancelled for medical reasons.

Prism prescription requirement all incorporated:

For distance:

OD +1.00 +0.50 × 90° 12Δ base-out 6Δ base-down = 20/25.

OS +0.50   13Δ base-out 6Δ base-up   = 20/20.

For near:

OD +3.50 +0.50 × 90° 6Δ base-out 6Δ base-down = J1.

OS +3.00   6Δ base-out 6Δ base-up   = J1.

Follow-up:

Six years six months (final office visit): wearing above perscriptions,

Titmus: 200 sec of arc.

Vectograph: fusion.

Worth 4-dot test: fusion at distance and near.

Maddox rod: E ½Δ   RH 6Δ,   E′ 1Δ   RH′ 6Δ.

NPC: good

"Able to drive again at night, prudently."

Thirteen years: patient contacted by phone. He is still comfortable with the same glasses and continues to see double without them.

Comments: This patient with a thyroid myopathy and a very long follow-up could have his diplopia corrected with Fresnel or incorporated prisms or both. However during two successive episodes, each with a follow-up of six years, he chose bilateral thick incorporated prisms over a single Fresnel membrane applied over his non-preferred eye because they afforded him better visual acuity and binocularity.

---------------------- Δ ----------------------

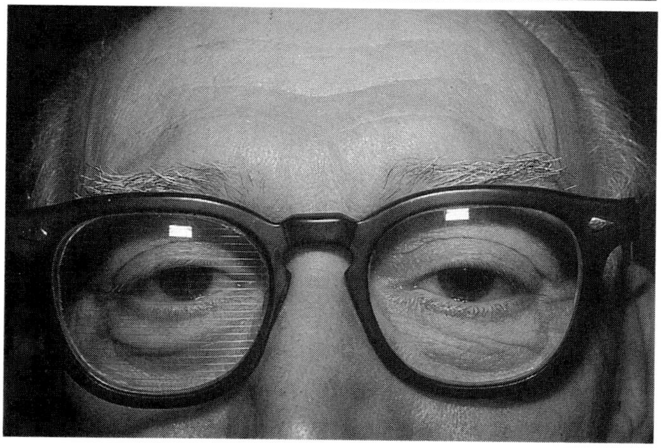

FIGURE 7-5. A, Patient L.L. with 12Δ base-out incorporated prism in his right lens and 13Δ base-out incorporated in his left lens (has worn 11Δ base-out incorporated in each lens for 6 years without diplopia). A 12Δ base-up Fresnel membrane has been applied over his left lens to correct a vertical strabismus of recent onset. He prefers higher power incorporated prism to any Fresnel membrane (see text). **B,** For near vision, patient wears 6Δ base-out over each eye. A 12Δ base-down Fresnel membrane has been applied over the lens of his right nonpreferred eye for near vision. The membrane will be replaced with incorporated prisms (see text).

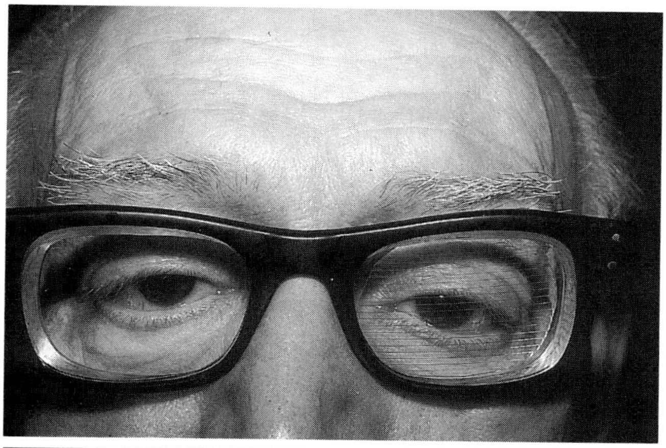

Orbital Surgery, Trauma, and Tumors

Surgical procedures, orbital trauma, and space-occupying lesions can affect the extraocular muscles and their innervations directly or can involve them indirectly from edema, congestion, and hemorrhage in the surrounding tissues. Spontaneous recovery is more frequently observed with indirect trauma. External scleral procedures, drainage devices used to control glaucoma, blepharoplasty with extensive fat resection, retrobulbar or peribulbar injection, orbital fractures and decompression procedures that can both induce scarring, and restrictive strabismus cause diplopia more frequently than paralysis from direct nerve involvement.

After implantation of drainage devices for glaucoma. Episcleral plastic drainage devices are being implanted in increasing numbers for the management of glaucoma refractory to conventional surgical procedures. Style and size of the implant, approach, and site of its placement are potential causes of motility disturbances.[31;32] They present as a restrictive strabismus with limitation of eye movements in the direction of the implant site. In the patient with long-standing stable or worsening diplopia, the management is difficult. Strabismus surgery on the involved eye takes the risk of compromising a functioning drainage device. Surgery on the fellow eye is reluctantly accepted by the patient.[32] The usually restricted visual fields complicate the correction of the diplopia by prisms. However, the highly motivated patient may adapt to a prism correction and may be further helped by combining the prism prescription with the use of prism to enlarge the visual fields (see Tubular Fields in Chapter 2).

After retinal surgery. Restrictive fibrosis has been implicated in the production of strabismus after all types of retinal surgery. In a prospective study Smiddy et al. found that scleral buckling can induce a muscle imbalance in the primary position in up to 23% of cases. The incidence of documented limitation of ductions was much higher.[36] If diplopia cannot be compensated with prisms, the presence not only of a cyclodeviation but also of aniseikonia should be ruled out. Several authors have reported a resolution of most of the primary position heterotropia within 6 months.[13;29] This would seem to be the minimum follow-up period. Follow-up should be prolonged if there is continuing improvement of the deviation and if the diplopia is corrected with prisms alone or with a mild compensatory head posture. Many authors have confirmed the benefit of prisms for treatment until resolution or stability is achieved.[13,42] (Case Presentation 7-6).

- **Case Presentation 7-6**

Patient L.V.: This 49-year-old man, with bilateral pseudophakia and no childhood history of strabismus, had horizontal and vertical diplopia in all fields of gaze following a scleral buckle for retinal detachment in the right eye 7 months earlier. The patient had been prescribed inverse vertical incorporated prism for distance (intentional ?) and was able to ignore the diplopia images. He had no prism in his reading glasses and constant, annoying diplopia at near. He did not wear a patch because he found readaptation too difficult. He avoided reading instead.

First exam:

Visual acuity (pc):   OD −3.00 −2.50 × 100° 5Δ base-in = 20/30.
OS −0.50 −0.25 × 130° 4Δ base-up = 20/25.
Reading glasses +1.50 OU = J1.

The patient had a right hypotropia of 14Δ increasing to 25Δ, in down and right gaze, with a small XT, greater at near. He had an excylotropia of 10° OD and 5° OS. He had a limitation of elevation of the right eye RIO > RSR and overaction of the depressors.

The findings were in favor of a posterior-inferior temporal mechanical limitation.

Office prism trial:

Diplopia overcome in the reading position at near with 30Δ Fresnel trial prism base at 75° over OD. Prolonged trial.

Prism prescription:

30Δ base at 75° Fresnel membrane applied to right lens of reading glasses.

Follow-up:

Ten weeks: patient was not helped at near with Fresnel prism. During the repeated office trial he experienced alternation of up and down vertical and torsional diplopia without fusion. The prism was replaced by a 0.1 Bangerter filter. Patient lost to follow-up.

One year five months (over 2 years after his retinal surgery): still experiencing diplopia when reading. Not comfortable with Bangerter filter. Wants to discuss surgery.

At examination patient was found to have a marked decrease of his vertical deviation, especially in down gaze, and only slight excyclophoria. With a 5Δ base-in prism he could compensate his residual right hypophoria.

Prescription:

OD −3.00 −2.50 × 105° 2Δ base-in (inc) = 20/30+.
OS −0.50 −0.50 × 30° 3Δ base-in (inc) = 20/20.
Add OD +2.00 = J2.
OS +2.00 = J1.

A letter from the patient 1 month later reported a remarkable improvement with the prism glasses.

Five-months follow-up: no diplopia at near and distance with prism bifocals.

Worth 4-dot test: fusion at distance and near.

Titmus: 80 sec of arc.

Maddox rod: LH 5Δ increasing to 9Δ in right gaze.

Comments: Although it is difficult to give the exact time when the patient began to spontaneously improve, it was probably at least 1 year after his retinal detachment surgery. It is also revealing that, even with the Bangerter filter reducing the vision to less than 0.1 in the right eye, this patient, whether the deviation was large or becoming smaller, still experienced such discomfort when reading that he requested surgery until his double vision was compensated by prisms. This was acheived only after his cyclotropia had nearly resolved, contrary to case presentation 7-3.

---------------------- Δ ----------------------

After orbital fractures. Another important group who illustrate well the different mechanisms involved in restrictive strabismus and benefit from prisms are patients with orbital fractures. Orbitomalar fractures are rarely accompanied by diplopia. Fractures of the floor of the orbit have the highest incidence of diplopia. Vertical diplopia with unpredictable evolution can be present even in the absence of detectable muscle entrapment or fracture.[28] It can also persist or appear after successful anatomic orbital floor repair.[5] For the latter two groups of patients and during the preoperative initial observation period, prisms can often control diplopia and should be prescribed after trial.

If limitation of both elevation and depression of the affected eye is present, prism spectacles that displace the reduced binocular field to a more useful space position may prove to be the treatment of choice (Case Presentation 7-7).

- **Case Presentation 7-7**

   Patient R.M.: This 36-year-old man was referred for diplopia, still present after his last procedure for orbital and facial repair for multiple fractures. The diplopia was vertical and constant in all fields of gaze. He had adopted a marked chin-up position to see in single vision, causing him pain in the neck.
   First exam:
   Visual acuity (sc): OD = 20/20,   OS = 20/20⁻.
   With chin up, patient fuses Worth 4-dot test at near and at distance.
   ACPT: RHT 25Δ XT 8Δ, RHT′ 14Δ XT′ 10Δ.
   Incomitant. Limitation of elevation left eye.

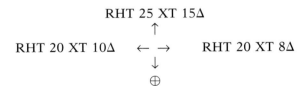

   RHT 25 XT 15Δ

   RHT 20 XT 10Δ   ← →   RHT 20 XT 8Δ

   ⊕

   Office prism trial:
   8Δ vertical minimum power required to correct chin position without diplopia, straight ahead, and in down gaze.
   Prism prescription:
   OD plano 4Δ base-down,
   OS −0.50 × 75° 4Δ base-up.
   Last follow-up:
   Eleven months: patient happy with prism glasses because of enlarged binocular field with single vision with head straight. He has returned to work. With prism correction he has vertical diplopia only in up gaze.

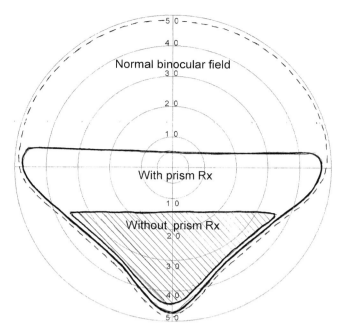

FIGURE 7-6. Binocular field of fixation of the patient R.M.. The interrupted line represents the limit of the normal field in a young individual. The crosshatched area delimits the field of this patient without prism, and the solid line represents the the improved field in up-gaze with the prescription of 8Δ vertical prism.

With head straight (ccΔ):
Cover-uncover test: flick RH, ⊕'.
Worth 4-dot test: fusion.
Stereoacuity at distance: 60 sec of arc (Vectograph). At near: 40 sec of arc (Titmus).
Deviation has become more comitant. An additional 6Δ vertical prism power could reduce the diplopia in up gaze without inducing diplopia in down gaze.
Surgical correction advised but patient undecided.

Comments: The binocular field of this patient (Fig. 7-6) demonstrates well that prisms, by extending his field just 5° above the midline, relieved him of his symptoms and allowed him to return to work. The patient is comfortable enough with his prescription that he hesitates to have surgery in order to discard it.

----------------------- Δ -----------------------

Diplopia Caused by Foveal Displacement

Epiretinal or subretinal neovascular membranes may cause a displacement of the fovea with resulting intractable binocular vertical or oblique diplopia.[3] The diplopia may be present only at distance. Unfortunately, it rarely responds to a prismatic correction. During the PAT, the vertical and the associated horizontal deviation, that may be small at the start, continue to build up without the examiner being able to find a prism that achieves stable fusion. These patients are not good candidates for prisms. In some patients, retinal stripping and laser applications have succeeded in removing the causative factor and curing the diplopia.

It has been postulated that a rivalry between central and peripheral fusion mechanisms is responsible. Even though a prism juxtaposes the diplopia images, stable single vision cannot be achieved. The stronger peripheral motor mechanism overrides the central sensory mechanism because of a slight eccentric mechanical distortion of the photoreceptor arrangement.[6] (Case Presentation 7-8).

- **Case Presentation 7-8**

Patient M.S.: This 51-year-old man had a 1-year history of vertical diplopia at distance of sudden onset, which was somewhat helped by raising his chin. Over the past 10 years he has had five episodes of central serous retinopathy in his left eye. A central fibrous metaplasia was documented by fluorescein angiography showing a serous detachment with inferior leak.

First exam:
Visual acuity (pc):   OD -2.25 $+1.25 \times 30°$ = 20/15, (sc) J1.
                                OS -2.25 $+1.25 \times 155°$ = 20/20, (sc) J2.
APCT (cc): LHT ½Δ, ⊕'.
Versions and ductions normal.
Maddox rod: LH ½Δ ①, LH' ½Δ ①'.
Amsler grid: superiorly displaced central scotoma OS.
Distance (Vectograph): fusion. No stereopsis.
Near stereoacuity (Titmus): 60 sec of arc.
Bagolini lenses:
   At near: fusion response.
   At distance: fusion with triplication of fixation light.
Afterimage test (twice):

Office prism trial:
Even with a prolonged trial, small amounts of prism would correct the diplopia only momentarily. LHT built up to 10Δ with slow recovery (1 week). Afterimage test repeated: suppresses OS.
Rx: none.

Comments: As with other published cases of binocular diplopia associated with retinal wrinkling and foveal displacement, *no amount of prism* could overcome the diplopia.[3;6] The lesion was also inferior with the affected eye slightly deviated upward. Although a prism buildup as seen in this patient has not been reported, it probably would have been observed with progressively stronger prisms.

The crossed diplopia with afterimage testing may have been induced by a nasal involvement of the fovea without horizontal diplopia in casual seeing. Horizontal diplopia is a rare sign of foveal displacement.

---------------------- Δ ----------------------

Diplopia After Aphakia or Pseudophakia in the Adult

No previous history of strabismus

Induced anisometropia and aniseikonia. In this decade when intraocular lenses, contact lenses, and corneal refractive surgery are the primary means for correction of aphakia, poorly fitted thick glasses are rarely the cause of diplopia after cataract surgery. However, the indiscriminate use of different correcting modalities or disparate powers in fellow eyes for bilateral aphakic corrections has resurrected the problem of diplopia induced by the sometimes severe anisometropia and aniseikonia. Young adults who have had unilateral aphakia corrected by a corneal refractive procedure or a contact lens in one eye only and who have had a posterior chamber lens placed later in the fellow eye or vice versa may complain of intractable diplopia secondary to aniseikonia. A difference of 6% or more in image sizes is generally not tolerated. Although patients with a marked degree of aniseikonia will notice immediately that the two images are of different size, in any aphakic patient, it is indicated to look for aniseikonia as the cause of a symptomatic diplopia that cannot be corrected with prisms. The Awaya test[1] based on image separation with red and green spectacles or the Ogle eikonometer are used to determine the presence and degree of the defect.

Frequently, a moderate anisometropia is created when one eye is made emmetropic and the other remains myopic or hyperopic. Such anisometropia may induce a decrease in stereoacuity, but diplopia, if present, is not persistent.

Decompensated heterophoria. The aphakic or pseudophakic patient with persistent diplopia may have a decompensated preexisting heterophoria or an unknown microtropia (see Case Presentation 7-10). Because the patient had no grossly apparent deviation before cataract surgery, a careful ocular motility evaluation that may have predicted this complication is not often done. Since cataract patients are being operated bilaterally at lesser intervals, a prolonged period of dissociation from the decreased vision in the unoperated eye is less likely to induce a strabismus. Nevertheless, horizontal diplopia, vertical diplopia, or both are still seen after cataract extraction.

In two series of cataract surgery done under general anesthesia and with the fixation suture placed in episclera anterior to the superior rectus, the incidence of diplopia in patients without previous history of strabismus was 1.9% (5 of 263 patients) in the first series[38] and 1.7% (7 of 412 patients) in the second series.[40] In all cases the diplopia resolved with time and the use of prisms. The patient may not have manifest diplopia but instead complains of blurry vision, which can be overcome by closing either eye. These patients can be very unhappy despite their technically successful cataract surgery. The examiner must be careful to rule out monocular diplopia. At times it is difficult to elicit any deviation with the cover test alone. The Maddox rod is an invaluable tool for allowing the examiner to elicit diagnostic responses.

The symptomatic patient, who has had a cataract extraction in one eye and a cataract in evolution in the fellow eye and who is not yet ready for surgery, is a good candidate for a prism. The prismatic correction necessary to overcome the diplopia can rarely be reduced before visual acuity is restored in the fellow eye. Following cataract extraction in the fellow eye, the prism power is decreased gradually. In most instances, improved fusional amplitude will allow the patient to discard the prism and remain asymptomatic, even though preoperatively the horizontal or vertical deviations may have been significant.

Secondary to muscle injury at the time of surgery. Devices pressed over the eye to lower the intraocular pressure before cataract surgery have been postulated to damage the inferior rectus by hypotoxic effect. Bridle fixation sutures applied to an extraocular muscle, usually the superior rectus, may cause hemorrhages and fibrosis. However, when the bridle suture is passed in the episclera anterior to the muscle insertion, it cannot be held responsible for vertical diplopia following cataract surgery. The next most frequent cause of diplopia following cataract surgery after decompensated deviations is a direct injury to an ocular muscle(s) during local anesthesia.

Retrobulbar anesthesia. With retrobulbar anesthesia the inferior rectus is the muscle most often injured in its posterior portion, deep in the orbital cavity.[17] On inspection at the time of strabismus surgery, *the peribulbar portion of the muscle may appear normal*. However, direct coronal and oblique sagittal views of the inferior rectus with orbital computed tomography (CT) or MRI has shown segmental enlargement of the inferior rectus in acute and subacute cases.[18] Segmental fibrosis resulted in the induction of a mechanical restriction. Because the latter is in the posterior part of the inferior rectus, the strabismus may be more marked in down gaze than in up gaze. When the adhesions are at the level of the anterior portion of the muscle or the inferior rectus is totally fibrotic, the hypodeviation is habitually more marked in up gaze. With either involvement a forced duction test will demonstrate a restricted elevation.

Another cause of the function deficit is the destruction of extraocular muscle fibers by direct exposure to myotoxic anesthetic agents. The quantity and concentration of the anesthetic solution accidentally injected into the muscle would be responsible.[11] Although it is likely that the muscle function will be restored by the regeneration of new muscle fibers, it may not be so in the elderly patient.[7]

Peribulbar anesthesia. This technique has recently gained favor as a safer procedure than retrobulbar anesthesia, especially to avoid accidental perforation of the globe. However, it too can induce strabismus and does so more frequently than retrobulbar anesthesia.[11] The local anesthetic solution is injected directly over the superior and inferior rectus muscles and may result in significant damage by myotoxicity or direct injury. A Brown's syndrome has been reported in one patient.[10]

A subconjunctival injection of antibiotics is frequently done at the conclusion of an anterior segment procedure. To induce less congestion postoperatively, some surgeons inject deeper in the sub-Tenons' space, injuring a muscle or provoking a local inflammatory reaction.

Patients who undergo ophthalmic procedures under peribulbar or retrobulbar anesthesia should be informed of the possible complication, although rare, of postoperative diplopia.

At the time of consultation, the majority of patients present with a strabismus from mechanical restriction. Forced ductions are mandatory for a final diagnosis. Primary and secondary deviations differ in paretic strabismus as well as in strabismus induced by mechanical limitation. The three-step test may be misleading, for example, a fibrotic superior rectus will respond as a homolateral superior oblique palsy. Although the surgical approach when necessary is different from that of a true paralytic strabismus, the prismatic approach is similar. Patients with bilateral intraocular lenses and good visual acuity may be reluctant to wear distance spectacles to correct their diplopia even if a small prismatic correction controls it. They should be encouraged to do so until they recover with time. Large deviations do not have such a good prognosis but should be handled with prisms during the observation period before surgery. These patients respond well to a recession procedure alone, preferably with an adjustable suture technique on the restricted muscle or combined with a resection of the direct antagonist. Contrary to paralysis of central origin these procedures can restore binocularity in all fields of gaze or leave only a slight vertical limitation (Case Presentation 7-9).

- **Case Presentation 7-9**

Patient A.H.: This 84-year-old man had a left inferior rectus contracture syndrome. Three months earlier he had undergone an extracapsular cataract extraction in the left eye with a posterior chamber intraocular lens under routine Van Lint and retrobulbar anesthesia [(4% lidocaine (Xylocaine) and hyaluronidase (Wydase)], 5 ml in each location. On removal of the eye patch he experienced vertical diplopia in all fields of gaze. He reported no prior history of strabismus.

First exam:
    Visual acuity (pc):   OD +6.00 = 20/30, J1.
                           OS +1.50 +1.00 × 109° = 20/50, J5.
                           Add +3.00 OU.

Incomitant.

At distance: left hypotropia of 12Δ in straight-ahead and up gazes, which increased to 18Δ to 20Δ in down gaze.

At near: with or without glasses, exotropia of 12Δ with hypotropia of 18Δ, which increased to 25Δ with XT of 8Δ in down gaze.

No cyclodeviation.

Ductions: marked underaction of elevation with mild overaction of depression in the left eye.

Prism prescription:

12Δ base-up Fresnel, upper segment OS (distance).

30Δ base-up Fresnel, lower segment OS (near).

Follow-up:

Two months (ccΔ): no diplopia straight ahead and in reading positions.

Worth 4-dot test: fusion distance and near.

Maddox rod: RH 2Δ ①, RH' $1\frac{1}{2}\Delta$ X' 8Δ.

Visual acuity: OS = 20/60, J7; (no Δ) 20/50, J5.

Prismatic correction cannot be reduced.

Comments: The greater Fresnel prism correction over the lower segment to compensate the diplopia at near was not caused by a base-down prism effect induced by his anisometropic correction. That complication had already been alleviated with a bifocal slab-off. The measurements were identical with and without glasses in the down position. At distance, as well as at near, the patient was more comfortable wearing the full prism power as is often seen with vertical strabismus of recent onset, especially in the presence of unequal vision and anisometropia.

---------------------- Δ ----------------------

Diplopia After Aphakia or Pseudophakia in the Adult

With previous history of strabismus. For the adult patient with decreased visual acuity that cannot be explained by the density of the cataract, a retinal pathology or the presence of undiagnosed amblyopia should be ruled out. A childhood history of strabismus is first suggestive of amblyopia. The patient should be warned not to expect better visual acuity after cataract extraction than that obtained with the Potential Acuity Meter or with an interferometer. In an amblyopia these instruments sometimes indicate a better potential acuity than that obtained after cataract removal.[12]

A patient with a previous history of strabismus will not be likely to see double if the dominant eye still remains the fixating eye after the cataract has been removed from the nondominant eye. However, if fixation is reversed after a cataract has been removed from the nondominant eye, diplopia may appear. Before such surgery is done the patient should be asked to hold fixation with the deviating eye. If the patient sees double and the diplopia cannot be overcome with prisms, intractable diplopia can be anticipated after cataract surgery. Muscle surgery will not be more successful than prisms to overcome the diplopia. If a cataract, even though less advanced, is also present in the dominant eye, it is indicated to operate on this eye first, though it is not always easy to have the patient accept to have surgery first on the eye with the better vision (Case Presentation 7-10).

- **Case Presentation 7-10**

  Patient D.H.: This 66-year-old male had intermittent diplopia and a marked ET following a cataract extraction OS 8 months earlier. He had noted an E(T) 6 years ago following his first cataract extraction OD. For the past 4 months he had been successfully wearing bilateral contact lenses. He denied any history of strabismus before his cataract surgery. The only photograph available showed a pseudoptosis with right hypotropia at age 35.

  First exam:
    Visual acuity (pc): OD (CL +14.50) with +1.00 × 10° = 20/20^{-2}.
    　　　　　　　　　　OS (CL +15.00) with −2.50 × 10° = 20/20^{-2}.
    　　　　　　　　　　　　　Add +3.00 OU = J1.
    ACPT (cc): LET 25Δ LHT 10Δ, LET′ 25Δ ⊖′.
    Afterimage test: alternate suppression.

  Prism prescription:
    After office trial with Fresnel trial set prisms to correct his diplopia, bilateral membranes of the same power and direction were applied to the patient's bifocal prescription, 12Δ base-out OD and 12Δ apex at 10° OS.

  Follow-up:
  One month, two months, and four months: deviation unchanged. With prisms, fusion with the Worth 4-dot test at distance and near. Asymptomatic.
    Titmus: 3000 sec of arc stereoacuity.
  Six months: needs only 12Δ base-out to correct his diplopia. Oblique prism removed OS.
  Eight months (cc no Δ): ET 10Δ, ET′ 8Δ.
    Prism OD reduced to 5Δ base-out.
  One year (cc noΔ): ET 8Δ ET′ 6Δ.
    Incorporated prism prescribed (over CL):
    OD −2.00 × 80° 2Δ base-out = 20/20.
    OS plano 3Δ base-out = 20/20.
    Add +2.00 OU = J1.
    Worth 4-dot test: fusion distance and near.
    Vectograph: fusion only.
    Titmus: 3000 sec of arc.
  Two years and eight months (ccΔ):
    ACPT: ⊕, X′ 1Δ to 2Δ.
    Maddox rod: E 4Δ LH 2Δ, E′ 1Δ LH′ 2Δ.
    Sensory state unchanged. Eye dominance changes with position of slit.
    4Δ test positive for microtropia OD.

Incorporated prism reduced to 1Δ base-out each eye. Minimum power for comfortable vision.

Six years: motor and sensory state unchanged.

ACPT (ccΔ): ⊕, ⊕'.

Amplitude of fusion: D 10 D' 8, C 14 C' 14.

Comments: This was a most cooperative patient. It is not usual that an adult will tolerate a 12Δ Fresnel membrane over *both* eyes for 6 months. When there is a strong eye preference it is indicated to favor that eye by applying the membrane over the fellow nonpreferred eye. This patient had equal vision in both eyes and a similar history of onset of esotropia with cataract development in the fellow eye and diplopia on removal of the cataract. His evolution and response to prismotherapy could have been that of a decompensated phoria. However, his binocularity, limited to peripheral fusion and gross stereopsis, his final absence of deviation with the alternate cover test, his lack of response to 4Δ base-out over his right eye, and the buildup of his deviation with "cataract occlusion" all point to the diagnosis of a decompensated microtropia.

Acquired unilateral cataract with secondary strabismus. Most such patients have a secondary exotropia. The age of the patient at the time of the cataract, the duration of the period without binocularity, and the quality of vision when the cataract is removed are all important factors, especially in the child.[39] For the adult, duration is not so important. Some adult patients, spontaneously or with a temporary prism, will recover normal stereoacuity even after being deprived from binocularity for many years. Therefore, after cataract removal, if the patient has diplopia, it is indicated first to prescribe prisms before considering muscle surgery (Case Presentation 7-11).

● **Case Presentation 7-11**

Patient M.H.: This 51-year-old man had aphakia OS, secondary XT, and constant diplopia after optical correction. At age 10 he had a traumatic cataract with onset of XT reported around age 30. Before his cataract was removed at age 38, visual acuity was reduced to light perception. Because of constant diplopia and poor tolerance to a corrective contact lens, the patient abandoned the effort to restore the vision of his left eye.

Exam:

Visual acuity: OD plano = 20/20.

                        OS +14.50 (trial CL) = 20/40.

ACPT (cc): LXT 35Δ RHT 4,   LXT' 40Δ RHT' 10Δ.

PAT (in office):

XT 30Δ, XT' 35Δ.

No diplopia, no vertical.

Surgery under microscope:
Recession LLR 7 mm.
Resection LMR 7 mm.
Follow-up:
Three months: Visual acuity (sc): OD 20/20.
(CL) OS 20/30.
(cc) flick LET, ⊕′.
Worth 4-dot test: uncrossed diplopia at distance, fusion at near.
Vectograph: uncrossed diplopia.
Titmus: no stereoacuity.
Seven months: (cc) orthophoric distance and near.
Stereoacuity: 120 sec of arc (Vectograph).
60 sec of arc (Titmus).
One year: sensory state unchanged.

Comments: This patient illustrates that fusion can be restored and stereoacuity can be regained after many years of binocular vision interruption, in this case over 20 years. PAT was essential to the surgical decision.

---------------------- Δ ----------------------

Diplopia After Occlusion in the Adult

Adult patients with amblyopia may complain of diplopia following full or part-time patching of their dominant eye, like cataract patients. This may even have been done on their own initiative to "exercise" their amblyopic eye. They erroneously apply a treatment to themselves, indicated only for the child, because of a fear that by using only one eye the other eye will become weaker. The adult amblyope should be cautioned against this practice. Prisms rarely help such patients. With time, if all patching is stopped, the patient will return to the old pattern of suppression (Case Presentation 7-12).

- **Case Presentation 7-12**

Patient T.M.: This 42-year-old man had sudden onset of intermittent horizontal diplopia in all fields of gaze for past 2 months. He reported an early onset esotropia with amblyopia left eye treated in childhood with patching and surgery about 4 years of age.
First exam:
Visual acuity (sc): OD 20/20 J1.
OS 20/40 J5.
Refraction: OD Plano = 20/20.
OS +2.25 +0.75 × 35° = 20/30.
APCT (cc): LET 4, LET′ 2.
Comitant.
Worth 4-dot test: ARC response distance and near.
Afterimage test: NRC.

Office prism trial:
With the trial of ½ hour, no amount of prism could eliminate diplopia. Careful questioning revealed a habit of closing the right eye to check the vision in the left eye. The patient was afraid that if he did not use the left eye the vision would decrease. The only treatment advised was to stop "self-occluding" his right eye.

Follow-up:
Two years: only occasional diplopia.
Flick LET, flick HT, flick XT'.
No stereoacuity.
Worth 4-dot test: ARC response distance and near.
Six years:
No diplopia. Onset of presbyopia.
Motor and sensory condition unchanged.
Prescription given.
OD −0.25 −0.50 × 80° = 20/20. Add +1.75 = J1.
OS +2.00 +0.75 × 35° = 20/30. Add +1.75 = J1.
Nine years:
Orthotropia.
4Δ test: positive for microtropia OS.

--------------------- Δ ---------------------

Summary and conclusions. The cases presented in this chapter illustrate well the importance of the sensory examination in the total work-up of the patient with an extraocular muscle dysfunction. Only by a thorough sensory evaluation can the examiner determine whether prisms are a necessary part of the total therapy. For example, in the last case presented, if a small prism had been used in an attempt to control the diplopia, it would have provoked an angle buildup and accentuated the problem. The indicated treatment, to stop the "self-occlusion," permitted the patient to return to his old pattern of foveal suppression, spontaneously relieving the diplopia. However, in case 7-10 in which a buildup to 25Δ ET and 10Δ HT had been provoked by "cataract occlusion" and for which strabismus surgery might have been done prematurely, prism therapy was the indicated treatment. This patient was returned to his monofixation pattern by gradually reducing the amount of prism that relieved his diplopia.

With the increasing age of the population, and with new techniques being used for local anesthesia, for glaucoma with episcleral drainage devices, for retinal surgery, for pseudophakia, and for refractive surgery, diplopia has become a more frequent complaint among adult referrals. Many of the cases presented illustrate well that prisms can help in the short- and long-term management of these problems, even in the presence of large angles and incomitance. Oblique Fresnel membrane prisms are very useful in the management of combined vertical and horizontal deviation, even in the presence of a cyclotropia, but the patient needs time to adapt. High-power incorporated prisms, preferred by some patients, are also an option.

REFERENCES

1. Awaya S, Sugawara M, Horibe F, Torii F: The "new aniseikonia test" and its clinical applications, *Acta Soc Ophthalmol (Japan)* 86:217, 1982.

2. Bever CT, Aquino AV, Penn AS, et al: Prognosis of ocular myasthenia, *Ann Neurol* 14:516-519, 1983.

3. Bixenman W, Joffe L: Binocular diplopia associated with retinal wrinkling, *J Pediatr Ophthalmol Strabismus* 21:215-219, 1984.

4. Bixenman WW: Vertical prisms. Why avoid them? *Survey Ophthalmol* 29:70-79, 1984.

5. Brabant P: Pre- and post-operative evaluation of diplopia and ocular motility in patients with orbital fractures, *Bull Soc Belge Ophtalmol* 224:111-122, 1987.

6. Burgess D, Roper-Hall G, Burde RM: Binocular diplopia associated with subretinal neovascular membranes, *Arch Ophthalmol* 98:311-317, 1980.

7. Carlson BM, Emerick S, Komorowski TE, Rainin EA, Shepard BM: Extraocular muscle regeneration in primates. Local anesthetic-induced lesions, *Ophthalmology* 99:582-589, 1992.

8. Dale RT: Indications for prismotherapy in strabismus, In Dale RT: *Fundamentals of ocular motility and strabismus*, New York, 1982, Grune & Stratton, p 393.

9. Diamond S: Prism management of vertical incomitance; Case reports; Conjugate prism correction for ocular myasthenia, *Tr Pac Coast Oto-Ophthalmol Soc* 46:135, 1965.

10. Erie JC: Acquired Brown's syndrome after peribulbar anesthesia (letter), *Am J Ophthalmol* 109:349-350, 1990.

11. Esswein MB, von Noorden GK: Paresis of a vertical rectus muscle after cataract extraction, *Am J Ophthalmol* 116:424-430, 1993.

12. Faulkner W: Laser interferometric prediction of postoperative visual acuity in patients with cataracts, *Am J Ophthalmol* 95:626-636, 1983.

13. Fison PN, Chignell AH: Diplopia after retinal detachment surgery, *Br J Ophthalmol* 71:521-525, 1987.

14. Greaves BP, Mein J, Gibb JC: Long-term follow-up of patients presenting with dysthyroid eye disease, In Moore S, Mein J, Stockbridge W, editors: *Orthoptics: past, present, future: transactions of the Third International Orthoptic Congress, Boston, 1975*, New York, 1976, Stratton Intercontinental Medical Book Corp, pp 223-240.

15. Guibor GP: Ophthalmic prisms; Some uses in ophthalmology, *Am J Ophthalmol* 26:833, 1943.

16. Guibor GP: Some uses of ophthalmic prisms, In Allen J, editor: *Strabismus Ophthalmic Symposium I*, St Louis, 1950, Mosby, pp 299-315.

17. Hamed LM, Mancuso A: Inferior rectus muscle contracture syndrome after retrobulbar anesthesia, *Ophthalmology* 98:1506-1512, 1991.

18. Hamed LM: Strabismus presenting after cataract surgery, *Ophthalmology* 98:247-52, 1991.

19. *Harrison's principles of internal medicine*, Editors: Wilson, Braunwald, Isselbacher, Petersdorf, Martin, Fauci, Root, ed 12, 1991, Chapter 316, Diseases of the thyroid, p 1703.

20. *Harrison's principles of internal medicine*, Editors: Wilson, Braunwald, Isselbacher, Petersdorf, Martin, Fauci, Root, ed 12, 1991, Chapter 366, Myasthenia gravis, pp 2118-2120.

21. Hart E: Case of extreme squint cured without operation by the use of prisms; with clinical remarks. A Mirror of the practice of Medicine and Surgery in the Hospitals of London. St Mary's Hospital (Ophthalmic Department), *Lancet* July, 30:119, 1864.

22. Huber A: Application of botulinum toxin in ophthalmology, *X Curso Internacional de Oftalmologia, (Barcelona)* Sept 18, 1993.

23. Hudson HL, Feldon SE: Late overcorrection of hypotropia in Grave's ophthalmopathy, *Ophthalmology* 99:356-360, 1992.

24. Lee J, Flynn JT: Bilateral superior

oblique palsies, *Br J Ophthalmol* 69:508-513, 1985.

25. London R, Scott S: Sensory fusion disruption syndrome, *J Am Optometr Assoc* 58:544-546, 1987.

26. Lueder GT, Scott WE, Kutsche PJ, Keech RV: Long-term results of adjustable suture surgery for strabismus secondary to thyroid ophthalmolopathy, *Ophthalmology* 99:993-1003, 1992.

27. Lyle TK: Management of ocular palsies of traumatic origin. Excluding reconstructive surgery in cases of fracture of the orbit, In Ferrer OM, editor: *Ocular motility*, Proceedings of the Horacio Ferrer Eye Institute Fifth Spring Meeting, vol 2, no 4, International Ophthalmology Clinics Boston, 1971, Little, Brown & Co, pp 146-176.

28. Lyon DB, Newman SA: Evidence of direct damage to extraocular muscles as a cause of diplopia following orbital trauma, *Ophthal Plast Reconstr Surg* 5(2):81-91, 1989.

29. Mets MB, Wendell ME, Gieser RG: Ocular deviation after retinal detachment surgery, *Am J Ophthalmol* 99:667-671, 1985.

30. Moore S, Welter P: Opthalmic diagnosis and evaluation of prism therapy for ocular myasthenia, *Am Orthop J* 43:97-101, 1993.

31. Munoz M, Parrish R: Hypertropia after implantation of a Molteno drainage device, *Am J Opthalmol* 113:98-100, 1992.

32. Munoz M, Parrish R: Strabismus following implantation of Baerveldt drainage devices, *Arch Opthalmol* 111:1096-1099, 1993.

33. von Noorden GK: *Binocular vision and ocular motility: theory and management of strabismus*, ed 4, St Louis, 1990, Mosby, p 391.

34. Pratt-Johnson JA, Tillson G: The loss of fusion in adults with intractable diplopia (central fusion disruption), *Austral New Zeal J Ophthalmol* 16:81-85, 1988.

35. Scott AB: Botulinum toxin injection of eye muscles to correct strabismus, *Trans Am Ophthalmol Soc* 74:734, 1981.

36. Smiddy WE, Loupe D, Michels, RE, Enger C, Glaser BM, deBustross S: Extraocular muscle imbalance after scleral buckling surgery, *Ophthalmology* 96:1485-1490, 1989.

37. Véronneau-Troutman S: Fresnel prism membrane in acquired extraocular muscle palsy, *Amer Orthop J* 24:91-97, 1974.

38. Véronneau-Troutman S: Muscle and fusion problems in aphakia, In Emery J, Paton D, editors: *Current concepts in cataract surgery*, St Louis, 1974, Mosby, pp 412-417.

39. Véronneau-Troutman S: Prognosis for binocularity in traumatic cataract, In Emery JM, Jacobson AC, editors: *Current concepts in cataract surgery*—Selected Proceedings of the Sixth Biennial Cataract Surgical Congress, St Louis, 1980, Mosby, pp 419-421.

40. Véronneau-Troutman S: Binocular disturbances associated with cataract evolution and correction, In Wisnia K, editor: CORNEA, Proceedings of XIV International Congress of SOBEVECO vol 14, Oostend, Belgium, 1986, pp 185-189.

41. Walsh and Hoyt's Clinical neuro-ophthalmology, ed 4, Miller NR, Baltimore, 1985, Williams & Wilkins, p 862.

42. Wenniger-Prick Maillette de Buy L, Van Mourick-Noordenbos A: Diplopia after retinal detachment surgery, *Documenta Ophthalmologica* 70:237-242, 1988.

GLOSSARY OF ABBREVIATIONS
USED IN TEXT AND CASE PRESENTATIONS

Heterophorias and Heterotropias

E	Esophoria at distance
E′	Esophoria at near
ET	Esotropia at distance
ET′	Esotropia at near
E(T)	Intermittent at distance
E(T)′	Intermittent at near
X	Exophoria at distance
X′	Exophoria at near
XT	Exotropia at distance
XT′	Exotropia at near
X(T)	Intermittent at distance
X(T)′	Intermittent at near
H	Hyperphoria at distance
H′	Hyperphoria at near
HT	Hypertropia at distance
HT′	Hypertropia at near
H(T)	Intermittent at distance
H(T)′	Intermittent at near
excyclo	Excyclodeviation (extorsion) in degrees
incyclo	Incyclodeviation (intorsion) in degrees
flick	Minimal $\leq 1\Delta$ or 2Δ
R	Add to above for right eye
L	Add to above for left eye
⊕	Orthophoria at distance, horizontal, and vertical
⊕′	Orthophoria at near, horizontal, and vertical
⊖	No vertical deviation at distance
⊖′	No vertical deviation at near
⦶	No horizontal deviation at distance
⦶′	No horizontal deviation at near
DVD	Dissociated vertical deviation
A pattern	ET greater in up gaze, XT greater in down gaze $\geq 10\Delta$
V pattern	ET greater in down gaze, XT greater in up gaze $\geq 15\Delta$
X, Y, λ,	Other alphabetic descriptive patterns

Vision and Refraction

OD	Right eye (oculus dexter)
OS	Left eye (oculus sinister)
FOD	Taking fixation with right eye
FOS	Taking fixation with left eye
OU	Both eyes (oculi uterque)
cc	With optical correction
pc	Present optical correction
CL	Contact lens
IOL	Intraocular lens
sc	Without optical correction
cc△	With optical and prism correction, incorporated or Fresnel membrane
cc no△	With optical correction and no prism

D	Diopter
Δ	Prism diopter
°	Degree sign: designates axis of a cylinder or orientation of an oblique prism on a circle, or a linear displacement: $1° = 2Δ$
+	Plus sphere or cylinder
−	Minus sphere or cylinder
x	Cylinder axis
Add	Bifocal or reading addition
CF	Counts fingers
LP	Light perception
J1 to J16	Near visual acuity, Jaeger designation
20/15 to 20/400	Distance with visual acuity, Snellen designation

Muscles

MR	Medial rectus
LR	Lateral rectus
SR	Superior rectus
IR	Inferior rectus
SO	Superior oblique
IO	Inferior oblique
R	Add to above for right eye
L	Add to above for left eye
Palsy	Includes paralysis with complete loss of function and paresis with incomplete loss of function

Tests

Cover-uncover	Unilateral cover test
ACT	Alternate cover test
ACPT	Alternate cover and prism test
NPA	Near point of accommodation
C	Convergence amplitude measured at 6 m
C′	Convergence amplitude measured at 33 cm
D	Divergence amplitude measured at 6 m
D′	Divergence amplitude measured at 33 cm
AC/A ratio	Ratio of accommodative convergence to accommodation
NPC	Near point of convergence
NRC	Normal retinal correspondence
ARC	Abnormal (anomalous) retinal correspondence
PAT	Prism adaptation test

Index

Styles of American Furniture

1860-1960

Eileen and Richard Dubrow

4880 Lower Valley Road, Atglen, PA 19310 USA

Library of Congress Cataloging-in-Publication Data

Dubrow, Eileen.
    Styles of American furniture 1860-1960/Eileen and
Richard Dubrow.
       p. cm.
    Includes bibliographical references and index.
    ISBN 0-7643-0157-8 (hard)
    1. Furniture--United States--History--19th century.
2. Furniture--United States--History--20th century.
I. Dubrow, Richard. II. Title.
NK2407.D82   1997
749.213--dc21   97-16558
                                CIP

Designed by "Sue"

ISBN: 0-7643-0157-8
Printed in China

Published by Schiffer Publishing Ltd.
4880 Lower Valley Road
Atglen, PA 19310
Phone: (610) 593-1777; Fax: (610) 593-2002
E-mail: Schifferbk.@aol.com
Please write for a free catalog.
This book may be purchased from the publisher.
Please include $3.95 for shipping.
Try your bookstore first.

We are interested in hearing from authors
with book ideas on related subjects.

Contents

Acknowledgements

There are many people we wish to thank for their help in putting together the information in this book. The list that follows is by no means complete, and we hope there is no one who feels purposely omitted:

Mr. J. Acunto
Jay Anderson
The Art Institute of Chicago
Art Museum, Priceton University
Baltimore Historical Society
Mr. Ralph Bloom
The Bostonian Society
Jay Brosig
Margaret Caldwell
Cathers Dumbrofsky
E. J. Canton
The Carnegie Instutute
Certified Auctioneers and Appraisers
The Chicago Architectural Foundation
Christie's East Auctions
The Chrysler Museum
The Cincinnati Art Galleries
Cincinnati Art Museum
Mr. and Mrs. J. Dobson
Wm. Doyle Gallery
Farm River Antiques
Crand Rapids Pub;lic Museum
Mr. and Mrs. H. Giontti
Barry Harwood at the Brooklyn Museum
The High Museum
David Hill Asian Art
Bill Holland
The Hudson River Museum
Inglett-Watson
Margot Johnson
Kerr's Antiques
Kurland-Zabar
Mr. Phil Lewis
The Lightner Museum
The Lockwood Mathews Mansion

Louisiana State Museum
Lyndhurst Corporation
Paul and Michelle Manganaro
George Mayer
Mr. & Mrs. T. Merlo
The Metropolitan Museum of Art
Herman Miller Archives
Ken Miller Auction
Mr. and Mrs. S. Mitchell
Norman Mizumo
Museum of Art, Carnegie Institute
The Museum of the City of New York
Musee des Arts Decoratif
Neal Auction Company
The Newark Museum
Peter-Roberts Antiques, Inc.
The Philadelphia Museum of Art
Post Road Antiques
Wm. Munson Proctor Institute
Public Museum of Grand Rapids
David Rago Arts and Crafts
Rick Rutledge at the Herman Miller Company
Savoiaa's Auction
Selkirk Galleries
The Shelburne Museum
Sotheby-Parke Bernet Auctions
The St. Louis Art Museum
Stingray-Hornsby
The Margaret B. Strong Museum
Treadway Gallery
University of California at Berkeley
The Virginia Museum of Fine Arts
Roberta Wagner
Western Reserve Historical Society
Mr. & Mrs. J. Wright

Preface

In our first book, *American Furniture of the Nineteenth Century* (1983), the revival styles in American furniture from 1840-1880 were presented to give an introduction to the many overlapping styles that were concurrently produced. Particular emphasis was given to American Rococo Revival furniture identification and cabinetmakers. There had been very little pictorial reference material previous to that work to help one understand the different styles, identify pieces from various cabinetmakers, or read a brief background of the major cabinetmakers.

This volume, in a sense, is a continuation of that study. However, this work emphasizes the dominant styles from 1860-1890, with an introduction to twentieth century pieces. Like the first book, it is a study for people learning about the styles, makers, and variations. This vast field is in its infancy of documentation, and we hope readers will help add to the knowledge. No single volume can present all the variations, but we feel the information should be recorded before it is lost altogether.

Wherever possible, chronological order has been maintained, although styles overlapped greatly and forms changed drastically under the intent of reform. The move toward more simplistic furniture in the twentieth century seems only an echo of the limited technologies (and therefore limited skills) of the seventeenth and eighteenth centuries.

We still struggle to come to terms with twentieth century design that has more in common with space technology than with the sitting room. While furniture design reaches out to grasp the benefits of the new technologies and materials, it simultaneously retains constant links with the past. We want our VCR, our television, and our stereo, but often we want them in "wood-grained" plastic cabinets. We want new wood furniture and then we add chipboard backs. We love the efficiency and durability of Formica for our kitchen table, but we want it with a wood-grain look; and we buy Formica for our office desk, shelving, and cabinets. In our living room, however, we seek old pieces or reproductions; in the bedroom we want Laura Ashley comfort.

It is interesting to note that while modern style has put a heavy hand on the design of public buildings, they are buildings in which the public has no voice. In the private sector, modernism has barely touched the average house. The same applies to furniture. While new furniture designs bring raves from the media, the average home reflects solid revival styles.

Chairs to the left and right are in the Gothic style; center chair by F.W. Krause of Chicago in the Arts and Crafts movement style, circa 1875. Courtesy Neal Alford Auction Company.

Bent willow rocking chair, circa 1900.
From the "Industrial Design" exhibit
at the Grand Palais, Paris. Courtesy
The Musee des Arts Decoratif.

Chapter 1
Historical Background

The Presidents of the United States
Between 1860 and 1960

1857-1861	James Buchanan	1893-1897	Grover Cleveland
1861-1865	Abraham Lincoln	1897-1901	William McKinley
1865-1869	Andrew Johnson	1901-1909	Theodore Roosevelt
1869-1877	Ulysses S. Grant	1909-1913	William Taft
1877-1881	Rutherford B. Hayes	1913-1921	Won Coolidge
1881	James Garfield	1929-1933	Herbert Hoover
1881-1885	Chester Arthur	1933-1945	Franklin Delano Roosevelt
1885-1889	Grover Cleveland	1945-1953	Harry Truman
1889-1893	Benjamin Harrison	1953-1961	Dwight D. Eisenhower

The Financial Developments

Until the Civil War, during the presidency of James Buchanan (1857), tariffs or customs duties constituted the main source of revenue for the United States federal government. The few internal revenue taxes in force had been largely abolished by 1817. Sale of public lands yielded large enough revenues so that the national debt was extinguished by 1835. However, the Mexican War created new debt and the Civil War created very serious financial problems. The treasury, after four long years of civil war, was forced to suspend redemption of greenbacks (legal tender notes) which depreciated severely in terms of their value relative to gold. Despite internal revenue taxes, heavy deficits were incurred.

After Abraham Lincoln became the President of the United States in 1861, the Morrill Tariff of 1862 gave public land as a subsidization of agriculture. This was followed by a series of war tariffs with duties that scaled upward and met the needs of Northern businesses. In short, politicians settled matters to benefit the financially strong East. The National Banking Act of 1863 authorized the establishment of national banks which could issue notes secured by U.S. Bonds. This Act caused the Independent Treasury System of 1846 to be disregarded in favor of private finance. The war-time taxes were repealed quite soon after the war, and customs duties resulted in budget surpluses.

In 1873, Congress eliminated the silver dollar and in 1879 special payment was resumed. During this time Congress also passed two Acts; the Bland-Allison Act and the Sherman Act, both of which allowed the United States to purchase silver with legal tender treasury notes redeemable in gold or silver. Neither act provided for unlimited coinage of silver, and therefore the monetary system was left on an essentially non metallic gold base. The Sherman Act was repealed in 1893 due to the reduction of the treasury's gold reserve. Following president McKinley's election, the gold standard Act was formally declared to consist of 25.8 grains, of gold 9/10 fine, and the Secretary of the Treasury was directed to maintain all forms of U.S. money at par. To insure a labor force, Congress in 1864 permitted the importation of contract labor from abroad. The act itself was repealed a few years later, but the practice continued until the 1880s. Contract labor provided a large labor force at low wages.

American growth in size and technology

During this same period nationally improved communications began. Telegraph and cable lines were subsidized, along with land grants to railroad promoters. The railroad subsidies meant easier access to the West. The Homestead Act of 1862 aided in populating the West. The Act enabled any citizen, except one who had served in the Confederate Army, to obtain 160 acres on the public domain by living on it or cultivating it for five years. With this new Act, the one billion acres in 1860 should have gone into the hands of individuals, however this was not the case. Only 1/6-1/10 of the public land was obtained by the homesteaders. The rest was not given away, but sold, or held off the market by speculators or the government. By the end of the century, it was figured that homesteaders had 80 million acres, the railroads had 180 million, the states had been given 140 million, and another 200 million (much of it Indian land) had been sold to the highest bidder.

While the homesteader policy seemed ideal, in reality it was never suited to the needs of landless working men or immigrants. It was nearly impossible to move a family west, build a house and farm, and buy equipment before any return came in for the year of hard work. The Fifteenth Amendment, which allowed U.S. citizens to vote regardless of "race, color or previous condition of servitude," helped to secure the black vote for the Republican party. In a similar way, the Fourteenth Amendment, which gave a constitutional guarantee of citizenship and equal civil rights to freedmen, was used by business to protect its interests from federal regulations. All of these factors, in addition to their claim of having saved the Union, kept Republican party politicians in power for a long time to come.

From rococo to practical furniture

In furniture design, the devastation of the Civil War seemed to bring an end to the exuberance reflected in the Rococo Revival furniture of the 1850s. Industry and commerce migrated from the Northeast to the more moderate Midwest, but the North maintained its prosperity. This was the "gilded age" of free spending for many, despite the fact that the years 1869 through 1890 were filled with strikes and economic depressions for the general population. A few wealthy industrialists, like Leland Stanford and Charles Crocker, lived in mansions; Andrew Carnegie lived very well in private railroad cars and castles in Scotland; and John D. Rockefeller lived in homes in Cleveland and New York. Most people lived in much simpler homes with practical furniture of moderate design and materials.

The 1880s and 1890s brought a great popular wave of protests against financial trusts and the so-called "robber barons". By the turn of the century, tycoons tended to retire to the Senate so that they could keep an eye on the country. Until 1913 state legislatures elected state representatives to the national Senate.

Evolving furniture styles

Throughout the period, changing economics and politics, combined with the growing technological revolution, brought an ever-increasing array of furniture styles. The Philadelphia Centennial in 1876 awakened American interest in decorative arts in the same way the Crystal Palace exhibition in London had done for the British in 1851. The Columbian Exposition of 1892 furthered that growing interest.

These upheavals caused interior designs to evolve from rich and exotic furniture to simpler Medieval, Mission, Art, and then streamlined modern styles. It was a period when materialism and science flowered and technology and expansion proceeded at a dizzying pace. While displaying wealth ostentatiously on the outside, the social problems multiplied and Americans in vast numbers questioned values from within. Gradually, Americans fulfilled their destiny as a nation of economic and social power, rather than mere supplier of raw goods to other nations. All the ingredients for the unparalleled material expansion of the next 100 years were in readiness.

Laminated wood furniture

John Henry Belter is often credited with inventing laminated wood furniture, but in truth laminating has probably been known since Egyptian times. It has been used by cabinetmakers throughout the ages.

In the 1770s in England, Thomas Chippendale used a three-ply mahogany for splats on dining chairs. In 1838, Samuel Craig of Boston patented an "elastic chair" made of continuous bentwood; its seat, rails, and front legs were made of a continuous piece.

Belter bent laminated wood in two directions in his "dishing chair," which provided not only strength but also a sinuous back curve of unusual grace. He patented this chair in 1858. Michael Thonet worked with laminated wood before he worked with bentwood. Isaac Cole patented a chair with a back, seat, and front support of one continuous piece in 1874. Gardner and Co. used 3-ply veneer for chair seats in their patents of 1874. This design continuum culminates in the twentieth century with laminated chairs designed by Eames, Danko, Rohde, and Alvar Alto.

Samuel Craig's "elastic chair," Boston, circa 1838. From the "Industrial Design" exhibit at the Grand Palais, Paris. Courtesy The Musee des Arts Decoratif.

John Henry Belter's "dish back" side chair, New York, circa 1858. From the "Industrial Design" exhibit at the Grand Palais, Paris. Courtesy The Musee des Arts Decoratif.

Platform rocking chair with perforated plywood seat and back, made by Gardner and Co. of New York, circa 1872. From the "Industrial Design" exhibit at the Grand Palais, Paris. Courtesy The Musee des Arts Decoratif.

Side chair (DMC chair) designed by Charles Eames, circa 1946. From the "Industrial Design" exhibit at the Grand Palais, Paris. Courtesy The Musee des Arts Decoratif.

Chapter 2
The Cabinetmakers in America

The following cabinetmakers were especially influential to changing styles of American furniture designs between 1860 and 1960.

Alvar Aalto (1898-1976)

Alvar Aalto, born in Kuorton, Finland, opened an architecture office in Jyväskylä, Finland, in 1923. In 1927 he moved to Turku and in 1933 to Helsinki. He began designing furniture during the 1920s, working independently and also creating pieces with his wife, Aino Marsio. The styles of the 1920s were based on classic design. Aalto made his birch stacking chairs in 1927 and his stacking stools in 1933. The technical innovation of these pieces was the placement of the seat on the legs in .such a way (with the legs extending to the sides) that the chairs could be stacked.

In 1930 he began his association with Artec, the firm that distributed and manufactured his furniture. The principle people involved in Artec were Aino and Alvar Aalto and Marie and Harry Gullichsen, who were considered friends and patrons of Aalto.

Aalto felt his most important furniture discovery to be the "bent knee," which was created by sawing a piece of solid birch open at the end, in the direction of the fibers, and gluing thin pieces of wood into the grooves. This idea was pioneered by Belter and Thonet and developed into modern design. Aalto's furniture found wide acceptance because it was light, practical, and reasonably priced.

His first wife, Aino, died in 1949. Alvar's second marriage was to a partner, Elissa Mckinieni, in 1952.

Joseph Albers (1888-1976)

Joseph Albers, born in Germany, was teaching at the Bauhaus when the Nazis closed it in 1933. He emigrated to the United States and taught at Black Mountain College in North Carolina. In 1950 he became chairman of the design department at Yale University. He is well known for his paintings as well as innovations in furniture designs.

Allen Brothers

The Allen brothers' firm was founded in 1835, in Philadelphia, by William Allen, Sr. (died 1869). His sons Joseph (born 1818) and William. Jr. were apprentices in their father's shop. In 1847, the sons took over the shop and the firm was listed in the Philadelphia City Directory as W & J Allen Furniture, 119 Spring Street, Philadelphia, Pa. William ran the business and managed the showroon while Joseph managed the factory. A third brother, James C., joined the firm in 1861. Then the firm was listed as W & J Allen and Brother. William, Jr. died in 1865 and his brothers continued as Allen and Brother until 1896. When Joseph died, James C. Allen took over and added as his partner Joseph A. Allen, Jr.

The firm won prizes from the Franklin Institute in 1854 and 1874, and received a commendation at the 1876 Centennial Exposition in Philadelphia.

Charles R. Ashbee (1863-1942)

Charles Robert Ashbee worked as an architect, industrial designer, and social reformer. He was the truest follower of William Morris's ideals of honest workmen producing work by hand, and believed that handicraft would restore dignity to the worker.

In 1888 he launched the Guild and School of Handicraft at Toynber Hall in London, England. The guild's most famous products were metalwork.

In 1900 he visited Chicago and met Frank Lloyd Wright, who later stayed with Ashbee in England. In 1902 he moved the guild to the rural Cotswold community of Chipping Camden. Due to the cost of bringing the guild product to market, the venture was liquidated in 1908.

Francis H. Bacon (1856-1940)

Francis H. Bacon originally was a draftsman at McKim, Mead, and White. Later he designed for Herter Brothers and also worked for H.H. Richardson. Later, he was the chief designer for A.H. Davenport and Company. After an unsuccessful bid to take over the Davenport company (between 1906 and 1914), he opened his own firm and remained in business until 1944.

Charles A. Boudoine (working 1829-1856)

Charles Boudoine was of French descent and went to France every year. (His wife Ann Postley was a milliner before their marriage). He imported French furniture, hardware and upholstery. His own furniture is a restrained, delicate interpretation of Rococo Revival style. He retired in 1856 and invested in real estate in New York. A designer for Boudoine was Mr. Kimbel later of the firm Bembe and Kimbel.

Berkey and Gay (J. Berkey 1861, working together 1866-1929)

In 1859, Julius Berkey was making furniture with James Eggleston; later he went into business with Alphonso Hamm, but neither partnership lasted. In 1861, Berkey alone opened a furniture shop on the second floor of his brother William's planing mill and sash and door works. In 1862, Julius Berkey and Elias Matter formed a partnership called Berkey and Matter. William bought a half interest in the firm in 1863 and the name became Berkey Brothers and Company and together they manufactured furniture, doors, and blinds. George W. Gay bought half of William's interest in 1866 and the name changed to Berkey Brothers and Gay.

Julius's business is best known as Berkey and Gay which was prosperous, for in 1880 the firm employed an average of 400 people. In 1886, David Kendall was their furniture designer, and in March of 1888, W. T. Wilson was the foreman of their factory. Julius Berkey and William R. Fox secured a patent in 1888 for furniture casters.

When George W. Gay died, his son Will H. Gay became head of the company which retained the name Berkey and Gay. Acquisitions enlarged the company in stages as it bought Oriel Cabinet Com-

pany in 1911, and the Wallace Furniture Company and Grand Rapids Upholstering Company in 1926. Then, in 1929, Berkey and Gay was sold to the Simmons Company of Chicago. A year later the firm discontinued production.

In 1935 there was an attempt to reopen the firm, and during the second World War it converted to war work, but post-war attempts to reorganize it were abandoned.

Marcel Bruer (1902-1981)

Marcel Bruer was born in Pecs, Hungary. At age 18 he went to Vienna, but left there shortly afterward to join the Bauhaus and Walter Gropius at Weimar. He remained at the Bauhaus from 1920 to 1928, first as a student and then as a master. In 1928 Bruer left the Bauhaus and moved to Berlin. He relocated to England in 1935. In 1937 he traveled to the United States to join Gropius in a partnership and to teach at Harvard University. This partnership dissolved in 1941; in 1946 Bruer left Harvard and moved to New York City.

In 1925, while at the Bauhaus, Bruer designed a tubular steel chair known as the Wassily armchair (named for painter Wassily Kandinsky). This was the first bent tubular steel chair of consequence. Tubular steel was a natural successor to bentwood, and would not have been a possible design without the earlier bentwood experiments of Thonet Brothers and J. and J. Kohn. Bruer is identified with the invention of the steel cantilever chair, which was introduced by Mart Stan and Ludwig Mies van der Rohe in 1927 at the Weissenhof Housing Exposition in Stuttgart (directed by Mies van der Rohe). The Bruer chair appeared at this exposition but was not manufactured until a year later. This 1928 cantilevered chair styled by Bruer and produced by Thonet became a classic of modern design. It was a better balanced and more practical chair than the Mies chair.

In 1935-1936 Bruer designed the lounge chair that was produced by Isokon, an English firm.

Bruer, along with van der Rohe and Le Corbusier, originated in the 1920s and 1930s the furniture that became popular in the 1950s.

Edward Colona

Edward Colona was born in Germany and worked in Dayton, Ohio, in the 1880s as a furniture and jewelry designer. Just before 1900 he moved to Paris to work for Samuel Bing (a native of Hamburg, Germany, who lived in Paris. Bing is credited with introducing the expression "Art Nouveau").

In 1895 H. Van de Velde of Brussels established the free design of the Art Nouveau movement in France.

Albert H. Davenport (working 1880-1906)

The firm A. H. Davenport and Company began in the facilities of Ezra H. Brabrook and Company, where Albert H. Davenport was the bookkeeper. The firm was located at 96-98 Washington Street, Boston, Massachusetts. Brabrook died in 1880 and Davenport purchased the firm.

In 1883 a building at 108 Cambridge Street in East Cambridge, Massachusetts, was purchased, and a branch was opened in New York City. The firm gained an international reputation. In 1882 and 1883, it decorated the Iolani Palace in Hawaii and some rooms of the White House for President Theodore Roosevelt. Davenport and Company also are known to have manufactured furniture for McKim, Mead, and White and H.H. Richardson.

After Davenport died, Francis H. Bacon tried unsuccessfully to buy the firm. By 1914 Irving and Casson were in control of the firm.

Donald Desky (1894-1989)

Donald Desky, born in Blue Earth, Minnesota, studied architecture and painting at the Ecole de la Grande Chaumiere, the Academie Colarossie, and the Atelier Leger in Paris from 1921 to 1923. On a trip through Europe he visited the Bauhaus and was very impressed with it and the Dutch De Stijl design. In 1927 he established the firm of Desky-Vollmer, with Phillip Vollmer, at 114 E. 32 Street in New York City. This firm was open until the 1930s. In the 1930s, Deskey designed for Ypsilanti Reed Furniture Co., and was one of the first Americans to design a line of chromium plated steel furniture. He also used Bakelite, aluminum, and neon lights to create decor with waterfall, Zodiac, and Mayan motifs. He had a greater regard for ornament and comfort than his Bauhaus contemporaries.

Charles Eames (1907-1979)

Charles Eames was born in Saint Louis, Missouri. In 1929 he traveled to Europe and returned to open his own office in Saint Louis in 1930. In 1936 Eliel Saarinen offered him a fellowship at Cranbrook Art Institute in Cranbrook, Illinois, which he accepted. Eames remained at Cranbrook for several years, and there met his wife, Ray Kaiser.

In 1940, Eames and Eero Saarinen won the Museum of Modern Art competition with a shell chair made of bent and laminated wood. Charles Eames

and his wife moved to California in 1941. Before the Second World War, Eames and Saarinen designs anticipated the postwar plastic chair. They reduced the number of parts in a chair and molded the plywood into a multicurved shell. Seat back and armrests were united in a single shell made of strips of veneer and glue laminated in a cast-iron mold. Eames made use of the strong, durable materials developed during the Second World War to create a one-piece chair that could be fabricated in one operation.

The future "Pre-Fabs" were his greatest contribution. His chairs were his best work in terms of material and design. He always attempted to create a product that could be mass produced. Eames is considered to be America's most important twentieth century furniture designer, and he proved that technology need not exclude art.

Charles Locke Eastlake (1836-1906)

Charles Locke Eastlake was born in Plymouth, England, and apprenticed as an architect to Philip Hardwick. He married in 1856 and traveled throughout Europe for three years. Eastlake never formally practiced architecture.

During the 1860s he published a number of articles on furniture, and then in 1868 published his *Hints on Holusehold Taste*. He suggested a style like Talbert, but it was more rustic and basic: he wanted simple functional design, honest construction, and the use of appropriate material and conventional ornament.

Eastlake designed wallpaper for Jeffrey and Company and furniture for Jackson and Graham. His strongest impact was as an arbiter of taste and as a challenger of the overt materialism and opulence of earlier centuries. The "Eastlake style" became popular in the United States after the death of William Morris in 1896, but this style was a free interpretation by mechanized furniture makers. American makers took Eastlake style to mean rectilinear form with construction of oak or walnut and the addition of gilt-incised surface decoration. In truth these pieces did not have much to do with Eastlake's principles.

Harvey Ellis (1852-1904)

Harvey Ellis was born in Rochester, New York, and attended the U.S. Military Academy at West Point. Little is known about his life except that he worked as a draftsman for H.H. Richardson while Richardson was a member of the architectural commission for the Albany City Hall. It is believed that Ellis later worked in the Midwest. In 1897 Ellis returned to Rochester and practiced architecture with his brother, Charles. The firm lasted five years.

Harvey Ellis exhibited at the Exhibition of 1900. He began to write for *The Craftsman* magazine and design furniture for the Craftsman Workshops of Gustav Stickley in 1902. C.F.A. Voysey and Charles Rennie Mackintosh were his greatest influences. Until Ellis joined the Stickley firm, they made no inlaid pieces.

George Grant Elmslie (1871-1952)

George Grant Elmslie was born in Scotland. He emigrated to Chicago in the 1880s and started as an errand boy in the office of Joseph Silsbee; by 1885 he was an apprentice. In 1890 he worked for Adler and Sullivan, and from 1894 to 1909 he was their chief draftsman.

In 1909 he joined William Gray Purcell (1880-1965) and George Fleich to form Purcell, Fleich, and Elmslie. Purcell and Fleich worked in Minneapolis while Elmslie worked in Chicago. Purcell and Elmslie were partners until 1922. Both men believed "the task of building is not that of a lone genius, but the coordinated effort of a number of men." The furniture they conceived was right-angle geometric with lyrical ornament, combining abstracted plant motif with linear pattern.

Wharton Esherich (1887-1970)

Wharton Esherich worked in Paoli (near Philadelphia), Pennsylvania, in the 1930s making free-form, unconventional, sculptural furniture. He developed a group of followers that included George Nakashima, Sam Maloof, and Arthur Carpenter.

Paul Frankel (1886-1958)

Paul Frankel was born in Vienna, Austria, in October of 1886. He received an architectural engineering degree in Berlin in 1911 and came to the United States in 1914. His work is allied with the Secessionist Movement. In 1922 he established a gallery at 4 East 48th Street in New York City where he sold his own and imported designs. His most famous designs are of skyscraper furniture, with the step back characteristic of a skyscraper. Along with Donald Deskey, he is considered one of the first American modernists in decorative arts.

Henry L. Fry (1808-1895) and William Fry (1830-1929)

Henry L. Fry and his son, William, were born in England. They were followers of the English Aesthetic Movement before emigrating to Cincinnati, Ohio, in 1851.

William Fry taught wood carving at the School of Design of the University of Cincinnati in the early 1870s. His son also taught there. Henry and William, together with Benjamin Pitman, are credited with the birth of the Cincinnati Art Furniture Movement and with the introduction of women in the furniture trade. Henry Fry's daughter was a wood carver.

Frank Furness (1839-1912)

Frank Furness was born in Philadelphia and was trained as an architect under Richard Morris Hunt. Furness served in the Civil War and then returned to Hunt's office to work. While his furniture stylistically does not seem to conform to the Arts and Crafts Movement, it does show the same rebellion against the mid-century revival styles. He designed pieces for specific interiors, and probably had his pieces manufactured in the Philadelphia cabinet shop of Daniel Pabst.

Thomas Godey

Thomas Godey was the successor to John Needles in Baltimore. In 1849 Needles and Son were at 54 Hanover Street, and Thomas Godey was at 54 1/2. Needles was not listed in the 1850 directory, but Godey appeared as the successor to John Needles. He was listed at 54 Hanover Street until 1858. He may have stayed in business until 1890, but at a different address. Godey manufactured furniture in the Gothic style.

Label for cabinet by Thomas Godey. Courtesy High Museum of Art, Atlanta Georgia. Gift of Virginia Carroll Crawford, 1981. 1000.41.

Eileen Gray (1879-1976)

Eileen Gray was born in Ireland and trained at the Slade School of Art in London. She served an apprenticeship at a lacquer workshop and then worked all over Europe with most of the great design names of the twentieth century. Gray was very much an individualist. Her most famous chair is the Transat chair (circa 1927) made for a house she built at Roque Brune in France.

Charles S. Greene (1868-1957) and Henry M. Greene (1870-1954)

The Greene brothers were born near Cincinnati, Ohio, but grew up in Saint Louis, Missouri. Charles studied woodworking at the Manual Training High School in Saint Louis. Both brothers graduated from Massachusetts Institute of Technology and in 1893 moved to Pasadena, California, to set up a practice in architecture.

They first were influenced by C.F.A. Voysey, then Frank Lloyd Wright, and then by Gustav Stickley and his magazine *The Craftsman*. Greene and Greene designs relate more to Charles Rennie Mackintosh. Charles loved wood, and his pieces reflect a passion for Oriental art.

Greene and Greene usually created for specific rooms and designed all of the pieces, even woodwork, fixtures, and so forth. They were true craftsmen, and even finished furniture themselves.

Walter Gropius (1883-1969)

Walter Gropius worked for Peter Behrens from 1908 to 1910. In 1914 he exhibited in Cologne, Germany, at the Deutsche Werkbund Exhibit. He joined the army for four years and then founded the Bauhaus in Weimar in 1919, and served as its first director. His work at the Bauhaus was very different from his earlier work. Gropius's early designs combined Biedermeier with classical detail.

He produced his first tubular steel chairs at the Bauhaus in 1925. In 1936 he left Germany to become a professor at Harvard University in Cambridge, Massachusetts, and was joined there by Marcel Bruer. Gropius revolutionized the training of designers. Students were taught to "search, probe, experiment, and seek a solution that would fulfill functional requirements and be the result of the material and tools employed." Invention and newness became of paramount importance to him, and the study of the past was disregarded.

A.S. Herendon and Company (1871-1885)

A.S. Herendon and Company was a dealer and manufacturer of furniture and mirrors in Cleveland, Ohio from 1871 to 1886. From 1871 to 1873 and 1876 to 1877, offices were at 114 Bank Street, Cleveland, Ohio. In 1885 offices were located at 189-197 Bank Street in Cleveland.

Herter Bros. brand. Courtesy Ken Miller Auction.

Herter Bros., Gustave Herter (1830-1898) and Christian Herter (1840-1883)

The Herter Brothers, Gustave and Christian, worked in a variety of styles and applied Eastlake principles and ideas.

Gustave was born in Stuttgart, Germany, and was adopted by the Herter family. He emigrated to the United States in 1848 and worked for Tiffany, Young, and Ellis (the predecessor of Tiffany and Company). By 1851 Gustave was listed as a cabinetmaker at 48 Mercer Street, and also was associated with the E.W. Hutchings firm at this time. In 1853 Gustave went into partnership with Auguste Pottier. For the New York Crystal Palace Exhibit of 1853-1854, he made pieces with Erastus Bulkley, who was his partner at 56 Beekman Street. (Another piece at the exhibit was made by Gustave in conjunction with T. Brooks.) Gustave moved to 547 Broadway in 1856 and remained there until he retired in 1870.

Christian Herter emigrated to the United States in 1860 and joined Gustave in the furniture business. The firm became known as Herter Brothers in 1865. From 1868 to 1870 Christian studied in Paris. When he returned to New York, he purchased Gustave's interest in the firm and Gustave retired. From 1870 until 1883 Christian Herter headed one of the foremost design and furniture firms in New York City.

After Christian died, the firm was continued under William Baumgarten until 1891 and then by William Nichols until 1906.

It is interesting to note that Wilhelm Kimbel (the son of A. Kimbel's brother, Martin), later of the firm of Kimbel and Cabus, worked for Herter Brothers.

Herts Brothers

The Herts Brothers worked at 806 Broadway in New York City and Broadway at 20th Street from 1872 to 1937. The firm designed at least one room of the Villard House in New York City and displayed a completely furnished bedroom suite at the Philadelphia Centennial Exposition in 1876.

Herts Bros. bedroom display at the Philadelphia Centennial Exhibition, 1876.

Bed by Herts Bros. displayed at the Philadelphia Centennial Exhibition, 1876. Birds-eye maple with mahogany carved birds, flowers, fruit and molding. The draperies were raw silk in blue and drab green with a canopy of tufted light blue silk.

Wolfgang Hoffman (1910-)

Wolfgang Hoffman, the son of Josef Hoffman, emigrated to the United States in 1925 and opened his own studio in 1927. Josef Hoffman, together with Koloman Moser, founded the Wiener Werkstätte to carry out theri belief that "materialism need not be crass, but could be refined and disciplined." Wolfgang's wife, Pola, designed textiles and executed pieces for him. The partnership, and marriage, dissolved in the 1930s.

Theodore Hofstatter, Jr.

Theodore Hofstatter, Jr., worked from 1871 to 1914 in New York City, and succeeded his father in the furniture business. He patented a reclining chair that is an American interpretation of the William Morris chair. Hofstatter received the patent for a simplified reclining mechanism. He was a furniture manufacturer and inventor of machines for furniture construction.

Elbert Hubbard (1856-1915)

Elbert Hubbard originally worked as a soap salesman. He visited William Morris in 1894 and, upon returning to the United States, established a community of craftsmen in East Aurora, New York. The community, known as "Roycroft," was modeled on some of Morris's principles, and started with bookbinding and printing before moving to handcrafted leatherwork and some furniture. Roycroft produced furniture that was basically "Work-a-Day" in style.

One of the community's most famous designers was Dard Hunter, whose Viennese origin influenced his work. In 1911 Hunter opened his own shop and school in East Aurora, New York.

George Hunzinger (1835-1898)

George Hunzinger was born in Tuttlinger, Germany, the eldest of four children in a cabinetmaking family. He emigrated to the United States and settled in Brooklyn, New York, in the 1850s. There he met and married Marie Susanne Grieb, also from Tuttlinger, Germany. He became an American citizen in 1865.

In 1866 he went into business for himself, as he already held a patent for a folding chair design. Additional patents and changes of his business location are indications of a successful career. Business locations included the following: 1866 - 192 Laurens Street, 1870 - 402 Bleeker Street, 1873 - 141 and 143 Seventh Avenue, 1877-79 - 3 temporary locations, 1879 - 323-327 West 16th Street.

The Hunzinger factory on Seventh Avenue was destroyed by fire on October 17, 1877. This fire also damaged Alexander Roux's establishment (which backed up to the Hunzinger facility). Hunzinger's business was conducted in three temporary locations until the new factory at 323-327 West Sixteenth Street was completed in 1879. The *Furniture Gazette* for June 1, 1886, noted: "George Hunzinger is still ahead with his rockers. They are original and in excellent taste and the workmanship and finish are all that can be wished."

In the late 1880s Hunzinger's sons, George, Jr. (born in 1862), and Alfred (born in 1868), entered their father's business. They tried to establish a foreign market for their furniture in additional to their American market.

After George Hunzinger, Sr.'s death, the executors of his will secured a patent of his in 1899. In the following years the firm made more conventional furniture, such as Morris chairs. In the twentieth century George's six daughters shared in the firm's ownership. Then, in 1920, rather than accept unionization of the factory or sale of the firm, the family closed the business.

Henry Iden

Henry Iden's firm was listed at 57 Elm Street, New York City in 1850. The company moved to 194-196 Hester Street in 1854. From 1855 through 1886 the address was 194 Hester Street, with the exception of 1866, when it was 196 Hester Street. In 1866 the firm's specialty changed from furniture to lamps and gas fixtures. In the 1860 directory a Henry Glisman is listed at the 194-196 Hester Street address as a furniture maker. On a bedroom set stenciled "Iden Cabinet Manuf.," with the address 194-196 Hester Street, Henry Glisman and Hensley are marked in pencil on the boards holding the mirror.

The Hensley listing found at 226 West 25th Street was noted as being in the gas burner business.

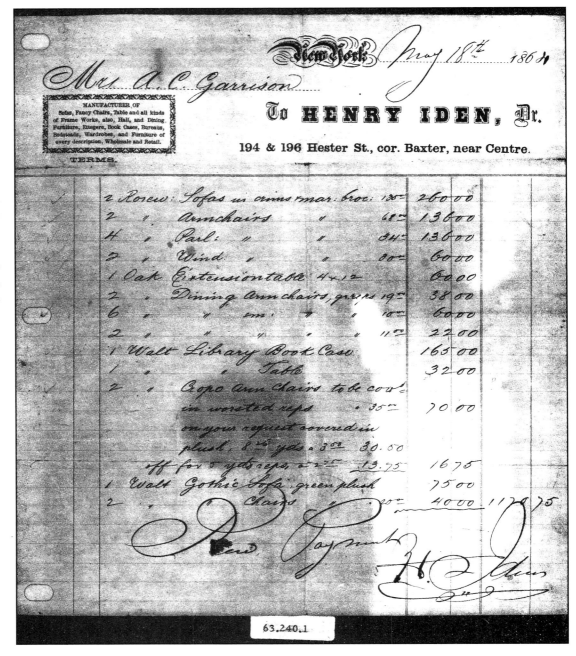

Bill of sale from Henry Iden, New York, 1864, for rosewood sofas and chairs, oak tables and chairs, and walnut book case, table, chairs and sofa.

John Jelliff (1813-1893)

At the age of fourteen, John Jelliff was apprenticed to Alonzo W. Anderson a carver in New York city. The arrangement did not work out and two years later he was apprenticed to Lemuel M. and Daniel B. Crane, cabinetmakers in Newark, New Jersey. In 1835, Jelliff was made a free man and in 1836 he went into business with Thomas L. Vantilburg. The partnership lasted until 1843 when Jelliff opened his own shop at 301-303 Broad Street. In 1860 Jelliff was forced to retire because of stomach trouble. Even after his retirement his firm was known as John Jelliff and Company but was run by Henry Miller who had been with Jelliff since he was a fourteen-year-old apprentice in 1843. He was married to the former Mary Marsh of Elizabeth, New Jersey. He worked in the tradition of the 18th century cabinetmaker and was an excellent carver.

Owen Jones (1809-1874)

Owen Jones, an architect by profession, was born in London in 1809. He designed the interior of the Crystal Palace in 1851 and published *The Grammar of Ornament* in 1856. He advocated flat, geometric ornament and the use of cast iron in architecture. As a theorist he influenced the entire design world of that time. He actually executed few buildings and is best known for his tile, carpet, silk, and wallpaper design.

Ely Kahn

Ely Kahn designed over thirty commercial structures in the zigzag style. He received his architecture degree from Columbia University and also studied at the Ecole des Beaux Arts.

Karpen Brothers

Solomon Karpen and his brother, Oscar, founded S. Karpen and Brothers in 1880. They were born in Posen, Germany, and emigrated to Chicago in 1872. Solomon, the oldest of nine brothers, began making upholstered furniture in a basement in 1880. In 1883 he and brother Oscar rented loft space to expand the operation. By 1887 they had a factory and a showroom, and their brothers, Adolph, Isaac, William, Benjamin, Michael, and Leopold, joined the firm. The youngest brother, Julius, joined the company in 1894. Isaac served as foreman of the upholstery shop, Oscar supervised frame making, Solomon purchased materials, William kept the books, Adolph was a salesman, Julius handled credit and correspondence, Leopold managed city trade, Michael controlled commercial trade, and Benjamin's job is uncertain.

The brothers constantly introduced new machinery and techniques. They controlled the patents for automatic tufting machinery. In 1900 two employees of the Karpen business, Streich and Rush, invented a spindle machine that was used by Karpen and manufactured by the Universal Carving Machine Company (set up by Karpen).

In 1900 S. Karpen and Brothers became the first wholesale upholstered furniture manufacturer to appeal directly to the consumer. They advertised heavily in magazines and offered a certificate of guarantee.

Karpen Art Nouveau parlor furniture was innovative both in design and construction. Among the Karpen Brothers designers over the years were Walter Teague, Eugene Schoen, Ilonka Karasz, Donald Desky, and Kem Weber.

Furniture by Kimbel and Cabus of New York displayed at the Philadelphia Centennial Exhibition, 1876. "The oak sideboard illustrated in our engraving is an excellent example of the best style of Jacobean furniture. The arched panels in the rear of the upper portion are beautifully sculptured in low-relief with natural objects, treated somewhat conventionally, but still following closely natural types. Growing wheat, wild-rose bushes in full bloom, the lily-of-the-valley, a bending oak-bough loaded with leaves, and other distinctive forms, are carved with much delicacy and faithfulness to type. They are molded with great skill, and there is nowhere anything like stiffness or angularity of treatment. The same thing may be said of the friezes of the upper and lower portions of the sideboard, carved to represent garlands of grapes and vine-leaves, in which we discern much grace and freedom of line, as well as softened contours." *Gems of the Centennial Exhibition*, p. 138.

Anthony Kimbel (working 1863-1882) and Joseph Cabus

Anthony Kimbel came to the United States in the 1840s and was associated initially with Charles Baudoine. Joseph Cabus worked with Alexander Roux at some point.

In 1854 A. Kimbel was a partner in Bembe and Kimbel at 56 Walker Street, New York City. This partnership lasted until 1862. In 1862 Kimbel joined Cabus in the Cabus Shop at 924 Broadway, and by 1863 the firm was listed as Kimbel and Cabus. From 1867 until 1874 the business was listed at 928 Broadway and 136 East 18th Street, New York City.

The Kimbel and Cabus firm favored ebonized furniture ornamented with incised gilt decoration. They used very simple forms, revealed construction, followed Eastlake principles, and favored the "Modern Gothic" style. Their designs display the influence of Bruce Talbert.

In 1882 Joseph Cabus moved to 506 West 41st Street, where he worked as a cabinetmaker until 1897. Anthony Kimbel and his two sons, Anthony Jr. and Henry, reorganized as A. Kimbel and Sons. After the elder Kimbel died, the sons continued as decorators until 1941.

Le Corbusier, Charles Edouard Jeanneret (1887-1965)

Charles Edouard Jeanneret was born in Chaux-de-Fonds, Switzerland. He designed and built his first house at age 17 and assumed the name "Le Corbusier" in 1923. In 1928, together with Charlotte Peuriand, he created his first important group of tables and chairs. This group consisted of reclining chairs, armchairs with pivoting backrests, and easy chairs. His furniture always was integrated with architecture. Le Corbusier wanted furnishings to be free of the stigma of "artistic" or "personal." He reduced furniture to three categories: 1) chairs for working, relaxing, and so forth that could be used in any setting; 2) multi-use tables; 3) shelves or cabinets that were standardized and could be hung or free-standing. His arm chairs had large cushions fitted into a metal framework, thereby shifting the skeleton of upholstered furniture from the inside to the outside of the piece. While his furniture was always elegant, it was often difficult to manufacture and sometimes uncomfortable. Le Corbusier rediscovered the bentwood chair and replaced the word "furniture" with the term "equipment."

Le Corbusier was a cousin of Pierre Jeanneret. Together they developed the earlier concrete experiments of Perret, Gropius, Mies van der Rohe, and Oud and made Americans reevaluate conventional styles of architecture and furniture.

Arthur Lasenby Liberty

Arthur Lasenby Liberty opened his retail shop on Regent Street in London in 1875 to sell silk from the East. The merchandise expanded so that, by 1890 and through 1910, Liberty and Co. made and sold the very best in Art Nouveau furniture, textiles, silver, jewelry, and pewter wares. Little is known about the early years of the company because records were destroyed in a fire. However, the firm employed independent designers and also had its own design studio. The Liberty Style they popularized is one of the best known phases of English Arts and Crafts style.

Raymond Loewy (1896-1983)

Raymond Loewy was a French-born pioneer in American design known for his streamline style. He did industrial designs but came out of the advertising and graphic arts field. He influenced packaging, logos and mass-produced products and gave aerodynamic design to the Studebacker car.

Ignatius Lutz (working 1844-1860)

Ignatius Lutz was an upholsterer in Philadelphia in 1850 and later expanded to making furniture through the 1870s. While he was considered unpopular with the trade, he dealt with some of Philadelphia's most prominent families. He went out of business in 1879.

Charles Rennie Mackintosh (1868-1928)

Charles Rennie Mackintosh was born in Scotland. He began practicing architecture for the firm of Honeyman and Keppie in 1889. In 1891 he won a student prize in Italy, and in 1892 began his career as a designer. He married Margaret MacDonald in 1900. Honeyman and Keppie became Honeyman, Keppie and Mackintosh in 1904, and in 1913 he resigned from the firm.

Mackintosh created the Glasgow School with Herbert MacNair and Margaret and Francis MacDonald. Herbert MacNair married Francis MacDonald.

Mackintosh was acknowledged to be brilliant and original in his furniture designs. He used furniture to define space and brought interior space to life. People did not look out of place in rooms with furnishings he designed. For Mackintosh, Art Nouveau style was an evolution of architecture and design.

A.H. Mackmurdo (1851-)

A. H. Mackmurdo founded the Century Guild, a cooperative furniture effort. He had a progressive attitude toward the aesthetics of his day, and sought his inspiration in Italian Renaissance and Queen Anne styles. He was opposed to commercial production. Mackmurdo is credited with the "tall chair" movement and anticipated and preceded the Art Nouveau movement by 20 years. Selwyn Image painted the panels of his furniture and Bernard Creswick made hinges for his furniture.

George Washington Maher (1864-1926)

George Washington Maher was born in Mill Creek, West Virginia. His family moved to New Albany, Indiana, and then to Chicago. At age 13 he apprenticed as an architect to Joseph Silsbee, and later worked for Adler and Sullivan. In 1888 he opened his own office and urged the rejection of Queen Anne and picturesque revival styles in favor of "original ideas."

Maher felt that "function should determine design" and developed a "motif rhythm theory." This theory indicated that geometric shape should be combined with a stylized floral form to create a dominant motif that should be rhythmically repeated throughout the interior and exterior to create visual harmony. His early furniture was more elaborate than his later Prairie School pieces. The floral motif he chose often was the honeysuckle.

Philip Maher (1894-1981)

Philip Maher was the son of George Washington Maher. During the 1920s he traveled abroad. The young Maher shared his father's belief in the value of ornament.

Sam Maloof (1916-)

Sam Maloof was a follower of Wharton Esherich near Philadelphia. His 1974 walnut rocker is of laminated construction. The wood grain always shows in his work.

Leon Marcotte

In 1848, Leon Marcotte came to New York from France where he had trained as an architect and he was listed in the 1849/50 and 1850/51 New York directories as an architect. The 1852 directory lists him under furniture at the same address as his father-in-law, Ringuet Le Prince, while the 1854 directory lists him as a decorator.

The Marcotte firm designed and manufactured furniture in New York but were particularly known for their imports.

Daniel F. Meader (1801-1877)

Daniel Fitch Meader was born in Baltimore, Maryland, on December 15, 1801. He participated in the War of 1812 as a powder boy on the privateer *Amelia*. Meader attended the Naval Institute of Pensacola in Florida as a member of the navy. Andrew Jackson, then head of the school, offered him a midshipman's commission, which he refused. In 1824 he traveled to Cincinnati and went into retail business with John Justice. In 1840 he invested in the Newport Manufacturing Company, but it was destroyed in a flood.

Together with Joseph Walter, he formed a partnership in the Steam Bureau Factory at Front and Smith Streets in 1846. In 1850 the business was listed at 19 Smith Street. The 1859 directory listed the firm as Johnson, Meader and Company. In the 1860s the company was listed at 43 West Second Street; in the 1870s the business was renamed Meader and Company and then Meader Furniture Company until Daniel Meader's death, when Joseph F. Meader, Daniel's son, became president.

Mark in oval border reading: D. F. Meader & Co. Furniture Manufacturers cor. Smith & Front Sts. Cincinnati, Ohio.

J. and J. Meeks (working 1797-1808)

The Meeks family name first appears in the New York city directory for 1797 with brothers Joseph and Edward Meeks at 59 Broad Street. By 1801, Joseph had set up his own business, separate from Edward.

By the 1820s, Joseph's sons, John and Joseph W. were in the business with him. Two other sons, Washington (listed as an upholsterer) and William H. (listed as a cabinetmaker) may also have been active in the business. When he was old enough John's son, John Jr. was also active in the firm.

By 1833, J. and J. Meeks had an organized system of sales outlets along the Atlantic coast which supplied an area from Boston to New Orleans according to Joseph Meeks' obituary. The system was carried out by various family members in various cities with warehouses to stock inventory.

In New Orleans, a warehouse was opened January 4, 1821 by Joseph W. Meeks "retailer, New York cabinet furniture" (vol I, *Artists of New Orleans*). Thereafter J. W. Meeks and Company is listed in New Orleans city directories at three addresses. Joseph's son Theodore appears in the directories around 1832 to take charge of the New Orleans branch while his father seems to have returned to New York. The 1838 directory lists the company as J. W. and T. Meeks & Company and Theodore's residence at 40 Hevia. In 1841 there are listings for three Meeks but no listing for a furniture store. In 1842 there are four Meeks listed but no listing for a furniture store. Now Theodore is listed as "lessee of the Verandah (hotel)".

In the 1830s, the Meeks firm in New York was a competitor of Duncan Phyfe, Allison and Lannuier. In the 1850s they were John Henry Belter's competitor. While not too many pieces by Meeks have shown up in the Hepplewhite, Sheraton or Regency styles, the base of the etagere in the well-publicized broadside done by the firm seems to be an example of the Adams style.

Joseph Meeks retired in 1834 and after his death in 1868 the remaining family members gave up the furniture business to devote themselves to their vast real estate holdings.

The *American Furniture Gazette* for December 1, 1882 had an article about "old Joseph Meeks" and tells this story: "On November 25, 1781 when the British fleet evacuated New York at the end of the American Revolution, he went down to the Battery to see the British fleet sail off and assisted in hoisting, under the orders of George Washington, the first American flag that ever floated over New York".

Mitchell and Rammelsberg

Robert Mitchell and Frederick Rammelsberg formed a furniture company in Cincinnati, Ohio, in 1847. Mitchell directed the general business and merchandising while Rammelsberg supervised the factory. In 1855 they opened a branch in Saint Louis, managed by Mitchell's brother, William.

Furniture by Mitchell & Rammelsberg of Cincinnati, Ohio, displayed at the Philadelphia Centennial Exhibition in 1876. *Gems of the Centennial Exhibition*, p. 141.

In another venture the company took over the defunct shop of J. & M. Flaherty in Kentucky in 1862 under the direction of another Mitchell brother, George, and the former head clerk for Flaherty, John Hoffman. This branch was called Mitchell and Hoffman and Company.

When Frederick Rammelsberg died, William Mitchell purchased the Rammelsberg interest in the Saint Louis store. In 1864 William sold half of his interest in Saint Louis to his brother, Robert, making Robert and William equal partners in Saint Louis. That year Robert also bought the Rammelsberg interest in the Cincinnati plant, making him sole owner. Here he continued to use the name Mitchell and Rammelsberg.

After the Civil War, Robert Mitchell went to New Orleans from Cincinnati. He founded the firm

Mitchell, Craig and Company. They took over the retail store of C. Flint and Jones on Royal Street. This partnership continued until 1870, when Robert Mitchell returned to Cincinnati and his brother, George, became manager of Mitchell, Craig and Company. A branch was opened at 103 and 105 Camp, New Orleans, as Mitchell and Rammelsberg Furniture Company-of New Orleans. At this same time George is listed as the agent until 1875 with Robert as clerk. In 1876 Robert is listed as salesman.

When the firm exhibited at the Philadelphia Centennial Fair in 1876 and in 1888, the name Rammelsberg was dropped. As Mitchell and Rammelsberg, the firm advertised extensively in New Orleans newspapers and listed themselves as dealers in all kinds of furniture.

Paper label "Manufactured by Mitchell & Rammelsberg Furniture Co. Cincinnati, O." Courtesy Louisiana State Museum.

Laszlo Moholy-Nagy

Laszlo Moholy-Nagy was born in Hungary. In 1937 he went to Chicago and founded the "New Bauhaus" which later became known as The School of Design. In 1949 this merged with the Illinois Institute of technology to become the Institute of Design.

In 1926 Nagy wrote, "The Bauhaus workshops are essentially laboratories in which prototypes of products suitable for mass production and typical of our times are carefully developed and constantly improved." The ideal Bauhaus object was a functional, unornamented form that could be mass produced using a minimum of material and labor.

William Morris (1834-1896)

William Morris was born in London and was trained as an architect in the office of George Edward Street, but he never practiced on his own. Morris founded the firm of Morris, Faulkner and

Company in 1861. His aim was to reform the field of domestic artifacts; he wanted to "escape" from industrialization and machine production. The firm Morris, Faulkner, and Company first showed its work at the International Exhibition in 1862, and lasted until 1875 when it became Morris and Company. Morris is known as a developer of taste and as a stimulator of public awareness. After his first experiments in furniture design for the Red Lion Square, his main interest was in wallpaper, textiles, and so forth. However, he supervised the Morris Company designs for furniture. Architect Philip Web (1831-1915) and Holman Hunt were actually responsible for the Morris Company furniture designs.

The Arts and Crafts Movement in furniture grew out of the ideas of William Morris and his approach to furniture. While Morris talked about art for all the people, his pieces actually were expensive and made for wealthy patrons.

George Nakashima (1913-1990)

George Nakashima was born in Spokane, Washington, of Japanese descent. He studied at the University of Washington and then went to the Massachusetts Institute of Technology. Nakashima traveled abroad to France, India, and Japan, and upon his return was interred with his family during the Second World War. After the war's end and his release, he moved to Pennsylvania.

Nakashima's furniture was handmade and characterized by his use of natural form for artistic advantage. He was a follower of Wharton Esherich. He based some chairs on traditional Windsor designs. He took simple, familiar forms and brought them to modern design. His motifs are Oriental with American form. His conoid table and chair draw on Shaker tradition. Nakashima's furniture retains simplicity while revealing sturdiness.

Daniel Pabst (1826-1910)

Daniel Pabst was born in Langenstein, Hesse-Darmstadt, Germany, on June 11, 1826. He emigrated to the United States and settled in Philadelphia by 1849. On July 4, 1854, he opened his own shop at 22 South Fourth Street. Sixteen years later, in 1870, he moved to 269 South Fifth Street and remained there until his retirement in 1896.

Mr. Pabst had strong feelings for Germany. He helped other immigrants, belonged to the German Society in Philadelphia, and wrote and published poetry in German. He married German-born Helena Gross and they had seven children, only three of whom, Emma, Laura, and William, survived to adulthood. The family lived in a row house at 264 South Fifth Street.

Pabst made only the finest furniture and he favored what was known as the "Modern Gothic" style. This was the style promoted by English reformers like Talbert and Dresser. He made furniture for the architect Frank Furness from original designs Furness provided. Mr. Pabst took his son William into the business in 1894. Although his last listing is in the 1896 directory, he made furniture for family and friends after his retirement and until his death.

Although advertisements by Pabst stressed "original design," his famous sideboard seems to have as its source the design book published by Henry Carey Baird.

Pape Brothers

Pape Brothers was established in Cincinnati in 1851 by Edward W. and Theodore Pape. The 1871-1877 directories list the firm as Pape Brothers and Kügeman, with Emil Kügeman.

The firm made gilt and rosewood molding, looking glasses, and picture frames. They also manufactured veneers imitating walnut, ash, rosewood, and silver gray maple. The business claimed to be nationally known.

Pottier and Stymus

Auguste Pottier (1823-1896) was born in Coulommiers, France. He went to New York City in 1847. By 1850 Pottier was a journeyman sculptor with the firm of E.H. Hutchings and Sons. In 1853 he became a partner of Gustave Herter in the firm of Herter, Pottier and Company at 48 Mercer Street. At some point during the 1850s, he was the general foreman for the firm of Rochefort and Skarren.

Cabinet by Pottier and Stymus of New York displayed at the Philadelphia Centennial Exhibition, 1876. *Gems of the Centennial Exhibition*, p. 134.

William Stymus was the foreman of the upholstery room at Rochefort and Skarren. Shortly before Mr. Rochefort's death (in 1857), Rochefort offered Auguste Pottier the business. The firm then became Pottier and Stymus.

In 1859 Pottier and Stymus listed the firm as, late, B.E. Rochefort with the address as 623 Broadway. In 1864 Pottier and Stymus moved to Mercer Street but kept their showroom at the Broadway

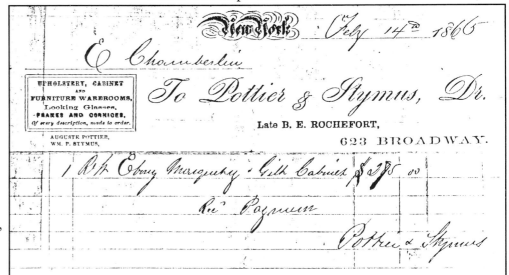

Bill of sale from Pottier & Stymus, New York, 1865, for an ebony marquetry and gilt cabinet.

address. They built a factory on Lexington Avenue and 41st Street in 1871. Between 1872 and 1882 they were listed at 325 Lexington Avenue.

The factory at 325 Lexington Avenue burned down in 1888, and Pottier and Stymus (as such) was liquidated. The company's stock was sold to a new cooperative whose officers included Auguste's nephew, Adrian, as president and William P. Stymus, Jr. The final listing for the business in the New York City directory was in 1918.

Victor Proetz (1897-1966)

Victor Proetz was born in Saint Louis, Missouri, and worked as an architect and designer. He graduated from the School of Architecture of the Illinois Institute of Technology in 1923. In 1929 Proetz won awards for his interior design of the Park Plaza Hotel in Saint Louis. He served as director of the Department of Interior Design of Lord and Taylor department store in New York City in 1944.

A. W. N. Pugin (1812-1852)

Augustus Welby Northmore Pugin was the prime developer of a progressive Gothic style in English architecture and design. His work inspired Isaac Scott's style in America. Pugin had two sons: E. W. Pugin (1834-1875) and Peter Paul Pugin (1851-1904).

Rammelsberg

See Robert Mitchell.

Henry H. Richardson (1838-1886)

H. H. Richardson was born in Boston. In 1860 he studied at Ecole des Beaux Arts in Paris. He returned to Boston in 1870 to set up his business. Richardson's designs relied upon the use of exposed natural materials that sometimes were carved, but there was little or no applied ornamentation. He was one of the first American architects to design furniture, and he realized that machines could create beauty if directed by artists. He brought together the best painters and sculptors he could fine, like John La Farge, Hunt and St. Gaudens. Richardson's work was influenced by the work of Shaw, Talbert, and Eastlake. There was great variety to Richardson's work, as he designed furniture with regard to a particular interior. His most famous commissions were for the Senate chamber of the New York State Capitol (for which he designed the desks); Crane Memorial Library in Quincy, Massachusetts; Unity Church in Springfield, Massachusetts; and Trinity Church in Boston (for which he became internationally famous). The A. H. Davenport firm in Boston executed his designs.

Gilbert Rohde (1894-1944)

Born in New York City, Gilbert Rohde's first jobs were in advertising. His furniture of the 1930s reflected the 1920s designs of M. Stam and Marcel Bruer. In the early 1930s he designed for Kroehler Manufacturing Company; in the late 1930s for Herman Miller. Until Rohde began working for Miller, the company made only reproduction pieces. Rohde was the first designer to put Bakelite tops on tables with chrome plated legs.

Rohde created multi-purpose furniture with interchangeable units. His designs suggested that the furnishings dictated the lifestyle of the owner. In keeping with the European Modern Movement, the furniture was efficient and socially equalizing. He was succeeded as a designer at Herman Miller by George Nelson. Rohde's various designs were produced by the manufacturers Heywood-Wakefield, Thonet Brothers, Kroehler Manufacturing Company, Herman Miller, Troy Sun Shade, John Widdicomb, and Mutual Sunset Co.

Ludwig Mies van der Rohe (1886-1969)

Ludwig Mies van der Rohe was born in Aachen, Germany, in 1886. In 1905 he apprenticed to furniture designer Bruno Paul, and later worked for architect Peter Behrens. In 1927 he was vice-president of the Deutscher Werkbund. Mies van der Rohe was largely responsible for the glass and steel buildings and furniture of 1920s Germany. The Barcelona Building (German Building) at the 1927 World's Fair was one of his best works. After 1930 he did not create any new works but continued to improve and develop his older designs. In 1930 he succeeded Walter Gropius as director of the Bauhaus and served as its final director (the Bauhaus was closed in 1933 by the Nazis). In 1938 he emigrated to the United States to join Walter Gropius, and served as director of the Architectural Institute in Chicago, which later became the Illinois Institute of technology.

Mies van der Rohe's greatest contribution was the skyscraper. His buildings were few but magnificent. His problem seemed to be the ease in which his designs were duplicated— and they were copied all over the world. Mies van der Rohe had an exceptional sense of proportion and craftsmanship.

Charles Rohlfs (1853-1936)

Charles Rohlfs was born in New York City and trained at Cooper Union. In the early 1870s he worked designing cast iron stoves. In the late 1870s he turned to acting with the Boston Theatre Stock Co. and became well-known. He married novelist Anna K. Greene and helped adapt her book *The Leavenworth Case* for the stage. He played the lead in it in the 1891-1892 season. Mrs. Rohlfs wanted him to give up acting so he turned to his hobby of woodworking. He opened his small shop in Buffalo, New York in 1890, and made single custom pieces.

Rohlfs' style was simple and functional. He decorated the strong outlines of his oak furniture with delicate carving and marquetry inlay. Like Gustav Stickley, Rohlfs relied on rectilinear forms and hand craftsmanship. Rohlfs was one of the few American designers who could design Art Nouveau pieces even though his early pieces had characteristics in common with "Mission style." Rohlfs always did the designing, while George Thiele often did the ornamentation. Rohlfs mark on furniture was an "R" within the rectangular frame of a wood screw. He retired in the mid-1920s.

Alexander Roux (1813-1898)

Alexander Roux was born in France and emigrated to New York City, where he married three times and had six children. He manufactured furniture in Gothic, Rococo, Renaissance, and Louis XV styles. In 1855 he employed 125 people. A set of Renaissance Revival bedroom furniture in the Henry Shaw House at 479-481 Broadway, Saint Louis, Missouri, made by Alexander Roux dates no later than 1866. This set shows how early he switched from the Rococo style. The height of his success occurred around 1870.

The American Furniture Gazette for December 1, 1882, related this about the house of A. Roux at Eighteenth Street in New York City: "It was a small house, packed so full of superb furniture, that either the customer or a salesman had to stand on the sidewalk when a purchase was being made."

Alexander Roux retired in 1881, but the firm continued under his son until 1898. Roux was cited for fine Louis XV furniture by Andrew Jackson Downing in *The Architecture of Country Houses.*

A. Roux and Company label including, at the bottom, Joseph Cabus's name as Alexander Roux's partner.

Alexander Roux label from a rosewood cabinet, reading: "Rouxs/Broadway/479/Meubles et Sieges Moderne/Furniture Warehouse/ No. 479 Broadway, New York." Circa 1856-1866. Courtesy High Museum of Art, Atlanta, Georgia. Virginia Carroll Crawford Collection. 1982. 311.

Label for A. Roux & Co. from a rosewood cabinet. Courtesy Farm River Antiques.

Stenciled label on wood reading "From A. Roux French Cabinetmaker as 479 & 481 Broad, New York."

Eero Saarinen (1910-1961)

Eero Saarinen, the son of architect Eliel Saarinen, was born in Finland. His family emigrated to the United States when he was 13. He went to Paris to study sculpture in 1929. In 1930 he attended the Yale School of Architecture, and graduated in 1934. Saarinen joined his father's practice and the faculty of Cranbrook Institute in Cranbrook Illinois, in 1936. In 1940 he won the Museum of Modern Art competition, along with Charles Eames, for a molded plywood chair they designed. In the late 1940s he designed the grasshopper chair and the womb chair for Knoll Associates. In the 1950s he created molded plastic pedestal furniture; in 1956 his colored and spun aluminum furniture became symbolic of the space age.

Eero Saarinen was very eclectic in his creativity. His largest contribution was bringing history, materials, color, and texture into architecture. He is considered the father of Post-Modernism.

Eliel Saarinen

Eliel Saarinen, the father of Eero Saarinen, came to the United States from Finland in 1924 as a visiting professor at the University of Michigan. He stayed on to head the Cranbrook Academy of Art.

George A. Schastey

George A. Schastey was a cabinetmaker and decorator working in New York City in the years between 1870 and 1879. He was listed in the New York City directory of 1869, but then was working for Pottier and Stymus. In 1873 he went into business for himself at 216 West 23rd Street. He moved to West 53rd Street in 1876, the same year his sideboard won a prize at the Philadelphia Centennial. In 1890 he moved his firm into the quarters vacated by Pottier and Stymus.

Schastey's work ranked with Herter Brothers, Pottier and Stymus, and others. He used laminating on the stretchers of some tables and very fine releif carving and inlay.

Oak sideboard by George A. Schastey of New York, 1876, with ash-root panels, about 16 feet wide and 16 feet high. *Gems of the Centennial Exhibition*, p. 140.

Eugene Schoen

Eugene Schoen was an American with an architecture degree from Columbia University. After seeing the Paris Exposition of 1925, he opened his own design shop at 115 E. 60th Street in New York City where he sold his own designs as well as imports. In 1929 he movedd to 43 West 39th Street. Karpen Brothers manufactured some of his designs.

Isaac E. Scott (1845-1920)

Isaac Scott was born in Manayunk (Philadelphia), Pennsylvania. He joined the Union Army when he was 17 years old, and after the Civil War became an artist and architect. Philadelphia directories of 1867 and 1869 listed Scott as a carver at 677 Bankson Street. He moved to Chicago sometime in 1872 or 1873. In the 1873 Chicago directory, he was listed as a designer at 149 S. Sangamon Street. His address in Chicago changed each year, but he was listed in the city until 1883. In 1875 he was commissioned by Mr. and Mrs. John J. Glessner to make pieces for their home. They remained his lifelong patrons.

Scott designed buildings and ceramics as well as furniture. In 1884 he went to New York City to practice interior design. He was listed in the New York City directories until 1888, at which point he moved to Boston to practice wood carving. He was listed in Boston directories until 1894.

Scott's style was based on the Gothic designs of Pugin in London, and it adhered to Eastlake principles. It was similar to the work of Herter Brothers. He was most successful in his designs for houses. Scott died in Melrose, Massachusetts.

Sligh Furniture Company

Founded by Charles R. Sligh, Sr., in 1880 at Grand Rapids, Michigan, Sligh Furniture Company was a furniture manufacturing plant. From 1894-1897 the firm also made bicycles. The company made furniture in walnut and later in mahogany. They used mirrors and glass from Germany until World War I when they switched to Belgian glass and mirrors. The firm was liquidated in 1932 due to the Depression and the effects of southern competition in the furniture market.

The Sligh Furniture Co. reorganized in 1933 under Charles Sligh, Jr. in partnership with O. W. Lowry at Holland, Michigan. They took over the failed Thompson Manufacturing Company. R.H. Macy's store in New York City purchased maple furniture made in the older style by the Sligh company and also by this company under the name Thompson Manufacturing Company. In the 1950s they were innovators of "crosscountry" furniture that was the standard of "ranch style" furniture of the period. In 1945 Sligh Furniture purchased the Grand Rapids Chair Company, and sold it in 1957 to Baker Furniture. In 1968 the Sligh Furniture Company purchased the Trend Clock Company.

Members of the Sligh family have been active in politics on a national and local level. John Brower was one of their furniture designers in the late 1890s. He created the first "modern" suites in Grand Rapids, furniture that was made with kiln-dried gum wood in a process that prevented warping and checking of the pieces. The furniture anticipated by almost twenty years the styles shown at the Exposition d'Art Decoratifs in Paris in 1925.

Gustav Stickley (1857-1942)

Gustav Stickley trained as a stone mason. He began making furniture with his uncle in the 1880s. The Gustav Stickley Company was founded in 1898 by Gustav and his four brothers in Eastwood, a suburb of Syracuse, New York. In the beginning the company made and sold ordinary walnut parlor furniture. During the mid-1890s he and partner Elgin Simonds manufactured reproduction Chippendale pieces. Stickley traveled to England in 1898, and upon his return began experimenting with his Craftsman Workshop Furniture designs. By late 1900, Gustav's brothers, Leopold and J. George, left him to form their own company in Fayetteville, New York. Another brother, Charles, was in business with Schuzler Brandt in Binghamton, New York as Stickley & Brandt.

It was in this period that Gustav Stickley's furniture took on the character for which it now is known. His designs were based on Charles F. A. Voysey's simple reform pieces, and were severely geometric. All traces of earlier decoration were absent: straight aprons were joined to the legs by mortis and tenon, and exposed pins held each joint. His furniture in this period was marked with a red decal that has a joiner's compass on it and the motto "als ik Kan" ("as well as I can").

Stickley's work took on a more Art Nouveau look in its decorative inlay when Harvey Ellis became his designer in 1902. In 1904, after Ellis's death, Stickley's creations grew increasingly simplified, and until 1909 were increasingly diversified. By 1910 this diversity disappeared and simplicity dominated. From 1910 until 1916, when his Craftsman Workshop went bankrupt, he clearly pointed the way to future furniture design.

The Arts and Crafts Movement ideal of "well-made and well-designed objects for the masses" was seldom obtained except by a man like Gustav Stickley and his furniture.

Leopold Stickley (1869-1957) and J. George Stickley (1871-1921)

Leopold and J. George Stickley left their brother Gustav's firm in 1900 to form their own business in Fayetteville, New York. Initially their designs were based on Gustav's work, but they did create unique pieces, such as the panel side settle. They seemed to be better at marketing their products than Gustav, and they maintained the quality and construction purity of the Arts and Crafts ideal. When Gustav went bankrupt in 1916, Leopold and J. George took over his workshop and renamed the firm Stickley Manufacturing Company.

Louis Sullivan (1856-1924)

Louis Sullivan was born in Boston and later attended Massachusetts Institute of Technology. In 1873 he moved to Philadelphia and worked for architects Frank Furness and George W. Hewitt. Later that same year he went to Chicago. In 1874 he traveled to Paris to study for a year at the Ecole des Beaux Arts. After holding a number of different jobs, he became a draftsman for Dankmar Adler, a structural engineer, in 1879. Adler contributed to architectural thinking significant concepts of structure, and believed that nothing was structurally impossible. In 1882 Sullivan became Adler's partner, and by 1883 the firm was renamed Adler and Sullivan. The partnership was dissolved in 1895.

Louis Sullivan continued to fight for new architecture until he died. He was the prophet of modern style and did much to emancipate architectural thinking from its conventional mode; his leadership paved the way for modernist architects such as Frank Lloyd Wright. Sullivan pursued skyscraper design by combining massive structures with intricate external detail. He attempted to have the building express the spirit of the activities carried on inside it.

H. H. Richardson's death in 1886 cleared the way for Louis Sullivan to produce the American counterpart of Art Nouveau. While Sullivan borrowed exterior forms from Richardson, he made them more severe and then enhanced them with ornamentation.

Bruce Talbert (1838-1881)

Bruce Talbert was one of the most influential industrial designers of the Aesthetic Movement and a leader in Gothic reform. His furniture shows straight lines, strap hinges, ring handles, architecturally inspired ornaments, revealed construction that shows dovetail and tenons, and was generally made in unstained oak. Born in London, Talbert mainly was an architect and designer. In 1867 he published *Gothic Forms* and in 1876, *Examples of Ancient and Modern Furniture*. Both publications were a plea for rationalism, solidity, and fitness of purpose in furniture.

Under the influence of Bruce Talbert and Charles Eastlake, neo-medieval furniture lost its fancy-dress look. Their designs opened the way for Arts and Crafts styles to develope.

Walter Teague (1883-1960)

Walter Dorwin Teague attended school at night while working as an advertising illustrator. For twenty years he continued in advetrtising before accepting his first industrial design commission for Kodak cameras in 1927. Some of Teague's designs were produced by Karpen Brothers.

Michael Thonet (1796-1871) and Thonet Brothers

Michael Thonet was born in Boppard-am-Rhine, Germany. As early as 1831 he was experimenting with the bending of wood but was making conventional furniture in his factory at Boppard-am-Rhine. By 1836 he was making laminated furniture with bent veneer pieces that were not particularly successful. In this period, his chairs were in the Biedermeier style and made of thin, narrow pieces of veneer that were stacked together, tied, and soaked in boiling glue. The bundles then were bent in wooden molds, dried, assembled, and thinly veneered to unify the design. In 1851 Thonet exhibited at the London Exposition.

In 1853 Thonet Brothers of Vienna was formed, having been named for Michael Thonet's five sons, under the protection of Prince Metternich. In 1856 the business was moved to the Black Forests of Moravia (Czechoslovakia) where they began using unskilled labor in the entire factory. Each laborer performed only one step in the process, giving the firm its mass-production capability. No firm of that day could match Thonet Brothers' volume of furniture production.

In 1860 Thonet introduced his rocking chair, followed by the swivel office chair and folding chairs. After Michael Thonet died, his sons continued to expand the business throughout the 1870s and 1880s. The New York City branch of the firm opened in 1874 at 826 Broadway to retail and finish furniture shipped in to it. In the 1890s the company invited designers Koloman Moser, Josef Hoffman, and Adolf Loos to design for them. In the 1920s the Thonet firm started manufacturing in the United States. In the 1920s and 1930s, Thonet Brothers manufactured the designs of architects Marcel Bruer (a cantilevered tubular steel chair that became a universal prototype), Mies van der Rohe, and Le Corbusier. In 1939

E M TIBBETTS,

MANUFACTURER AND DEALER IN

FURNITURE

OF ALL KINDS,

COFFINS, CASKETS AND ROBES;

CHILDREN'S CARRIAGES.

Agent for American Express Co.   DEXTER, ME.

Trade card for E. M. Tibbetts, undertaker and furniture maker, Dexter, Maine, front and back.

the firm moved its headquarters to New York City and it continues in business today.

The Thonet company achieved the Arts and Crafts ideal in a union of technology and aesthetics that enabled the production of inexpensive, handsome products for the masses.

Louis Comfort Tiffany (1848-1933)

Louis Comfort Tiffany was the artistic son of Charles Lewis Tiffany, the founder of the Tiffany and Company retail store in New York City. At age 18 he was a student of painter George Iness, and at age 21 he went to Europe to study painting under Leon Bailly. Louis Tiffany and Samuel Coleman traveled to Africa before Tiffany returned to the United States in 1870 and opened his own studio. But he returned to Paris in 1875, gave up painting, and turned to decorative art work and developed "favrile glass." He perfected many techniques that made his work with glass superb and original. Tiffany returned to the United States in 1879 and, in association with Samuel Coleman, Lockwood DeForest, and Candice Wheeler, formed Associated Artists. In 1900 he established Tiffany Studios in New York. Samuel Bing of Paris became his European agent after seeing the Tiffany display at the Columbian World Exposition. Louis Tiffany was influenced by the Eastern art he witnessed and became a leading exponent of the American Aesthetic Movement.

Charles Toby (1831-1888) and the Toby Furniture Company

Charles Toby was born in Denis, Massachusetts. He left Cape Cod as a young man to work in a Boston furniture store. In 1855 he went west, and while on the train, met an agent of the Boston furniture company and arranged to represent the firm in Chicago. The Boston business failed in one year, and

Toby was left with the store and promissory notes. He borrowed $1000 and started his own business. In 1856 his brother, Francis, joined the business as a partner. The firm survived the panic of 1857 and the three-year depression that followed. They then prospered from the economic upswing of the Civil War.

In 1865 they joined with F. Porter Thayer to form Thayer and Toby Furniture Manufacturers. This company survived the Chicago fire of 1871 and reopened immediately. The firm made horn furniture as a ready-made line in 1883 and, most importantly, had a special order department. In 1875 the brothers bought out Thayer, and Frank Toby took over daily operation of the factory. By 1880 Frank Toby opened departments to sell tiles, wallpaper, and drapes, and he vastly expanded the special order department. The company introduced model apartments, already furnished, for inspection and selection of furniture in 1886.

In 1889 Frank Toby formed a partnership with Wilhelm F. Christiansin, who had been a superintendent with John A. Colby and Sons. The new partnership aggressively advertised, marketed, labeled, and promoted its handmade furniture. By 1905 they had a store in New York City on West 32nd Street that sold only Toby handmade furniture. They also were enterprising marketers of factory made Arts and Crafts Movement furniture.

Joseph Urban

Joseph Urban was born in Austria and had trained at the Berlin Academy of Applied Arts. He came to the United States in 1914. Between 1922 and 1924, he was the manager of the Viennese Design Workshop in New York, a branch of the Wiener Werkstätte. He worked in the skyscraper style and believed that furniture in the apartment should harmonize with the outside skyline. He designed interiors for the Boston Opera House and the Metropolitan (New York) Opera House and furniture for the Mallin Furniture Company.

Gotlieb Vollmer (-1883)

Gotlieb Vollmer was born in Württemberg, Germany, and came to America in 1830. His first partner in business was David Klauder in 1843. Between 1845 and 1854, Xavier Montre was his partner, and thereafter he worked with his son, Charles.

Charles Vollmer had been sent to France to learn the furniture business, which began as an upholstery shop and expanded to include cabinetmaking. They worked in the tradition of hand crafting.

One of the Vollmer's clients was President James Buchanan, who bought a set of furniture for his home, "Wheatlands," in Lancaster, Pennsylvania, and another set for the Blue Room of the White House in Washington, D.C.

Wardrobe by G. Vollmer and Co. of Philadelphia, 1876. *Gems of the Centennial Exhibition*, p. 73.

Furniture by G. Vollmer and Co. of Philadelphia, 1876. *Gems of the Centennial Exhibition*, p. 144.

Charles F. A. Voysey (1957-1941)

Charles F. A. Voysey was born in Yorkshire, England. In 1874 he was apprenticed to J. P. Seldon, who had been a cabinetmaker before becoming an architect. Voysey then worked with George Devey. He opened his own office and designed fabrics and wallpapers, both of which were to become internationally known, in 1882. In 1890 he began designing furniture, and his work was a fusion of Art Nouveau and Arts and Crafts. It almost always was made in oak with plain surfaces and complex metal hinges.

Voysey was a designer, not a craftsman, and he did not execute his designs. He was interested in the utility of a piece and in eliminating ornament; A. H. Mackmurdo was Voysey's single greatest influence. Voysey designed wall paper and fabrics, and was a founding member of the Art Worker's Guild with his best designs, as well as his best houses, being done between 1895 and 1911. He pioneered a style that became known as Art Nouveau.

George E. Ware

In *Stimpson's* (of Boston) *1860-1861 Directory*, George Ware was listed as a member of the firm of Blake, Ware, and Company, furniture dealers. In 1862 he formed his own firm with an address of 12 Cornhill and 25 Washington Street in Boston. The company was listed until 1875.

Kem Weber (1889-1963)

Karl Emanuel Martin (Kem) Weber was born in Germany and apprenticed in 1904 to Eduard Schultz before being a journeyman under Bruno Paul. He went to San Francisco in 1914 to assist in work on the German Pavillion at the Panama-Pacific Exposition (1915) and was trapped there when war broke out in Europe. After the armistice, he opened a studio in Santa Barbara, California. Three years later, he went to Los Angeles to work as a draftsman for Barker Bros. In 1927 he opened another studio "to make the practical more beautiful, and the beautiful more practical." Karpen Brothers in Chicago produced some of his furniture designs.

Frank Lloyd Wright (1869-1959)

Frank Lloyd Wright was born in Richland Center, Wisconsin. He attended the University of Wisconsin, studied civil engineering, and graduated in 1888. He began his practice of arcitecture in the office of Joseph Silsbee and then worked for Adler and Sullivan. In 1894 he opened his own office in Chicago and designed furniture as well as houses. His early efforts, like the Robie House, have low-pitched roofs and horizontal thrusts. His early furniture was made by John W. Ayers; later Wright used Matthews Brothers Furniture Company in Milwaukee to execute his designs. Gillien Woodwork Corporation succeeded Matthew in manufacturing his designs. In 1904 Wright designed metal office chairs and furniture for the Larkin Company in Buffalo, New York. In the period between 1916 and 1920, Wright used the floating cantilever in construction of the Imperial Hotel in Tokyo, Japan. Japan's 1923 earthquake justified Wright's use of this construction technique. "Taliesin West" in Phoenix, Arizona, and "Falling Water" in Bear Run, Pennsylvania, are among Wright's best known examples.

Frank Lloyd Wright was concerned with every aspect and detail of his buildings, even hinges and fixtures. He disliked painting and sculpture so much that plants and window walls replaced space that could be used to show other people's art. Wright was in revolt against the formal pattern of architecture, and while one could say there is a "Le Corbusier chair," "table", or "cabinet," with a Wright design you only would choose it as a solution to a particular artistic problem. His furniture and houses show spartan simplicity and progressive ideas and reflect an admiration for Japanese culture. The furniture often was built-in and its construction was not visible.

At his homes, Wright trained a select group of architects in his principles of pure design derived from abstract geometric principles. In 1955 he created three lines of domestic furniture for Heritage Henredon Furniture Company of North Carolina.

Chapter 3
Renaissance Revival (1860–1890)

The years 1860 to 1890 were characterized by the style known as Renaissance Revival, which encompassed at least four different phases.

The first phase drew on sixteenth century French furniture designs that gave an angular appearance to the piece and terminated in elaborate pediments, baroque cartouches, carved animals, and human figures. The preferred wood was walnut, just as it had been in the sixteenth century. Elegance was added to the piece by the addition of marquetry, ormolu, and/or porcelain mounts.

In the second phase, the motifs seemed to merge with Louis XVI Revival styling from eighteenth century France. The furniture was characterized by gilt and ebonizing. Chairs and sofas had square backs with ear-like projections on their crests. The legs of the pieces were turned and fluted, and the decoration was either ormolu mounts or gilded incised lines.

The third phase is referred to as "Neo-Grec" or "Egyptian Revival" because it drew on classical and Egyptian motifs but not usually on Roman or Egyptian furniture form. The furniture took on a more two-dimensional look, with less carving but more walnut burl panels and inlays of fruit, flowers, animals, and musical instruments. Motifs such as the sphinx, lotus blossoms, and winged sun discs appeared after the Egyptian exhibit of 1852 and after the opening of the Suez Canal. Some pieces even used the klismos or curule leg to heighten the feel of antiquity.

The fourth phase encompasses the mass-produced furniture, much of it made in Grand Rapids, Michigan. Grand Rapids was ideally situated near large supplies of lumber; good water transportation; and a large, hard-working population including German immigrants who had cabinetmaking skills. A huge industry manufacturing furniture for the growing American population developed here and has continued to the present time. Renaissance Revival and factory production were ideally suited to one another. The factories could turn out these walnut incised styles in the quantities required by a growing middle class that could now afford them. The furniture turned out by these factories used scrolled pediments rather than carved ones and shallow carving done by shapers with very little hand finishing. This was not true of all "Grand Rapids" firms, since Mitchell and Rammelsberg, Meader, Phoenix, Nelson and Matter, Berkey and Gay, and a host of others also made exceptional pieces.

Rare tête-à-tête by John Jelliff of Newark, New Jersey. Courtesy Mr. and Mrs. Hertzberg.

Upholstered burl walnut arm chair and matching side chair with bronze medallions and gilt incising, attributed to Alexander Roux. Courtesy Yolanda and Jay Wright.

Suite of upholstered sofa and chairs by Schrenkeisen, patented 1875. Label on the rocking chair reads: Schrenkeisen's rocker Pat. Oct 20, 1874, Pat. May 23, 1875, N.Y. Courtesy George Wagner and Roberta Mayer.

Parlor of the Julia Ward Howe Home, 1900, showing a Renaissance Revival interior. Courtesy The Bostonian Society, Boston, Massachusetts.

Walnut Neo-Grec armchair with colored inlaid medallions and front marquetry panel. Courtesy E. J. Canton.

Very fine Revival ebonized and veneered sofa with applied fire gilt mounts. Courtesy E. J. Canton.

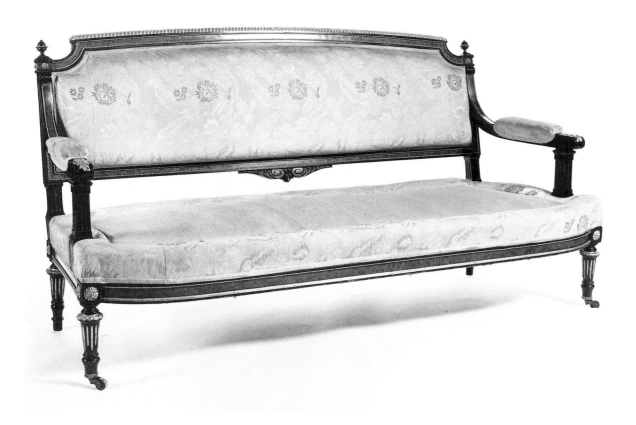

Mahogany Neo-Classical style armchair in the grand Beaux Arts American Renaissance tradition, and detail of arm and support. Courtesy E. J. Canton.

High style rosewood Renaissance Revival upholstered armchair with finely carved caryatid heads, with detail of the carving. Courtesy E. J. Canton.

Bentwood office chair made in 1876 by the Heywood Wakefield Company, walnut and burl walnut. This is a larger chair than their usual office chair. Courtesy Richard and Eileen Dubrow.

Ebonized side chair frame, part of a three-piece parlor set. Courtesy George Mayer and Roberta Wagner.

Upholstered tête-à-tête. Courtesy E. J. Canton.

Sofa, armchair, and side chair frames bearing the stenciled label of
A. Roux. Courtesy Farm River Antiques.

Armchair by Gotlieb Vollmer of Philadelphia, 1875. Courtesy Philadelphia Museum of Art, Philadelphia, Pennsylvania. Given by the heirs of Mr. and Mrs. James Dobson.

Grain-painted patented folding chair with original "carpet" upholstery by E. W. Vaill, Worcester, Massachusetts. Courtesy E. J. Canton.

Armchair attributed to Gotlieb Vollmer of Philadelphia, circa 1865. Ebonized cherry and twentieth (?) silk damask upholstery. Courtesy Philadelphia Museum of Art, Philadelphia, Pennsylvania. Given by the heirs of Mr. and Mrs. James Dobson.

Sofa by Leon Marcotte. Dimensions: 72" long. Courtesy Metropolitan Museum of Art, New York. Gift of Mrs. D. Chester Noyes, 1968.

Armchair and side chair.

Rosewood side chair from a parlor set, 1852, by Charles A. Baudoine. Proctor Collection. Courtesy Munson-Williams-Proctor Institute, Utica, New York.

Pair of walnut side chairs with burl walnut, incised gilt trim, and bronze medallions attributed to A. Roux. Collection of Mr. and Mrs. J. Wright.

Library (Metamorphic) chair designed by Augustus Eliaers, Boston, circa 1855. From the "Industrial Design" exhibit at the Grand Palais, Paris. Courtesy The Musee des Arts Decoratif.

Folding arm chair, circa 1860. From the "Industrial Design" exhibit at the Grand Palais, Paris. Courtesy The Musee des Arts Decoratif.

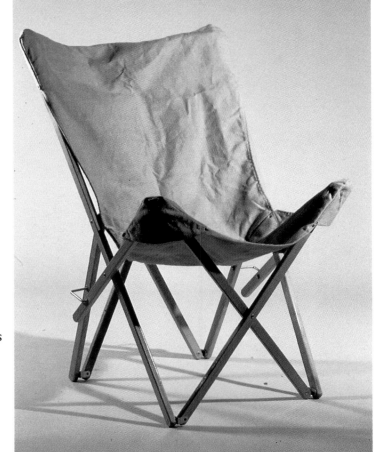

Sling chair. From the "Industrial Design" exhibit at the Grand Palais, Paris. Courtesy The Musee des Arts Decoratif.

42 Renaissance Revival

Armchair, two side chairs, and settee by John Jelliff, Newark, New Jersey, circa 1875. Renaissance Revival with Louis XVI and Neo-Grec influence. Carved rosewood female caryatids support the armrests, Athenian finials and ribbon framing cameo porcelain plaques. (Matching loveseat in the Centennial Collection, Smithsonian Institution, Washington, D.C.) Photo courtesy Stingray Hornsby Antiques, Watertown, Connecticut.

Renaissance Revival walnut side chair with ebonized, gold-incised details and bronze bird medallion. Original needlepoint seat. Dimensions: 34 3/4" high. Collection of Norman Mizuno Fine Art and Antiques.

Ebonized folding chair with gilt incising, gilt angle mounts, and original upholstery. Dimensions: 48 3/4" high. Collection of Norman Mizuno Fine Art and Antiques.

Renaissance Revival rosewood lady's armchair with carved bird's heads. Dimensions: 35 1/4" high. Collection of Norman Mizuno Fine Arts and Antiques.

Pair of Renaissance Revival ladies' chairs.

Chair back sofa and three chairs from a Renaissance Revival parlor set. Courtesy Neal Alford Auction Company.

Comb-back Windsor writing arm chair, 19th century. From the "Industrial Design" exhibit at the Grand Palais, Paris. Courtesy The Musee des Arts Decoratif.

Platform rocking chair made by Heywood Brothers and Co., circa 1873. From the "Industrial Design" exhibit at the Grand Palais, Paris. Courtesy The Musee des Arts Decoratif.

Invalid's chair manufactured by Marks A. F. Chair Co., circa 1876.
From the "Industrial Design" exhibit at the Grand Palais, Paris. Courtesy The Musee des Arts Decoratif.

Horn arm chair, circa 1885. From the "Industrial Design" exhibit at the Grand Palais, Paris. Courtesy The Musee des Arts Decoratif.

Armchair. Courtesy Maryland Historical Society, Baltimore, Maryland.

Pair of rosewood Renaissance Revival armchairs. Richly carved and inlaid, they feature bronze doré classical goddesses and exceptional porcelain roundels. A rare example of the collaborative work of Gustave Herter and August Pottier during their partnership in New York from 1853- 1859. Commissioned for the Rathbone family. Courtesy Stingray Hornsby Antiques, Larchmont, Connecticut.

High style Louis XVI Revival ebonized center table with applied fire gilt mounts. Probably made in New York. Courtesy E. J. Canton.

Walnut and ebonized Renaissance Revival centennial occasional table with marquetry top depicting George Washington, a bald eagle, and the dates 1776 and 1876, made by Kilian Brothers, New York City, circa 1876. Dimensions: 30" high, 28 1/2" wide 18" deep. An illustration from *Kimball's Book of Designs* (1876) shows an identical table. Photo by Dick Goodbody, courtesy Kurland-Zabar, New York.

Library table from the shop of Leon Marcotte. American. 49 7/8" long. Courtesy Metropolitan Museum of Art, New York. Gift of Mrs. Robert W. DeForest, 1934.

Renaissance Revival inlaid and ebonized table with brass figural mounts in the Louis XV style. Courtesy Richard and Eileen Dubrow.

Marble top Renaissance Revival table. Courtesy Western Reserve Historical Society, Cleveland, Ohio.

Rosewood center table decorated with bronze doré roundel plaques by Pottier and Stymus, New York City, circa 1870. Rare fossilized marble top framed by copper embossed surround depicting Roman warriors, hunters, and farmers. Mounts signed "P&S." Documented from the Cornelia M. Stewart House, 1 West 43rd Street, New York. Courtesy Stingray Hornsby Antiques, Watertown, Connecticut.

Rosewood card table (open) by Alexander Roux, New York City, circa 1850-1857. Courtesy High Museum of Art, Atlanta, Georgia. Crawford Collection. 1987.1000.43.

Center table attributed to L. Marcotte featuring ebonized wood with gilt incising, marble top, and bronze mounts. Courtesy Post Road Gallery, Larchmont, New York.

Rosewood center table with rosewood veneer burl walnut veneer, ormolu mounts, porcelain and metal plaques, and marble top. Attributed to Pottier and Stymus, New York, circa 1865-1875. Dimensions: 30" high, 46 1/2" wide, and 34 1/8" deep. Courtesy Lyndhurst Corporation, New York.

Walnut Renaissance Revival center table with burl maple border, gilt incising, and white marble inset. Dimensions: 32 1/4" high. Collection of Norman Mizuno Fine Art and Antiques.

(See more tables in color on page 65)

American black walnut desk (closed view) by Frank Furness of Philadelphia, circa 1875. Dimensions: 71" x 62 1/2" x 32 1/2". Courtesy Philadelphia Museum of Art, Philadelphia, Pennsylvania. Given by George Wood Furness. 74.224.1.

Desk by Brunner and Moore.

Fine rosewood secretary/bookcase. Courtesy Mr. and Mrs. T. Merlo.

Secretary by T. Brooks.

Secretary of walnut and burl walnut with gilt incising. Courtesy Farm River Antiques.

(See more desks in color on page 68)

Sofa and two chairs from an Egyptian Revival parlor set. Photos courtesy Neal Alford Auction Company.

Below:
Egyptian Revival settee with gilt incised, ebonized walnut and rosewood inlay. Bronze dore mounts depict sphinxes, lion heads, and Greek masks. The legs have stylized Pharoah's beard design terminating in hoofs. The mounts are signed P&S for Pottier and Stymus, New York, c. 1865.

Egyptian Revival armchair and settee, circa 1870.
Courtesy Margaret Woodbury Strong Museum,
Rochester, New York. Settee frame photo courtesy
of E. J. Canton.

Pair of Renaissance Revival upholstered lounges made of carved, incised, and gilded walnut. Late nineteenth century. Courtesy Lightner Museum, St. Augustine, Florida.

Neo-Grec lounge of carved, stained, inlaid, and turned wood, circa 1870. From the Gross Mansion, Chicago, Illinois. Courtesy Lightner Museum, St. Augustine, Florida.

Man's armchair from an Egyptian Revival set of walnut, burl maple borders, and gold incising. Bronze Egyptian heads with washed gold and other bronze mounts are at the sides. 39 1/2" high. Courtesy Norman Mizuno Fine Art and Antiques.

Egyptian Revival rosewood side chair with ash and prickly juniper veneer, possibly Pottier and Stymus Manufacturing Company, New York, circa 1870-1875. Dimensions: 28 1/8" high, 25" wide, 21 3/4" deep. Courtesy Metropolitan Museum of Art, New York. Funds from Various Donors, 1970.

Labeled Mitchell and Rammelsberg Renaissance revival arm chair with distinct Egyptian features in the crossed base and rolled arms. Courtesy Louisiana State Museum.

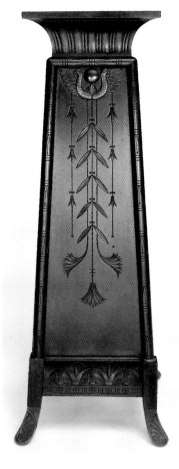

Egyptian Revival pedestal of carved and incised wood with bronze bird head mounts and feet. Late nineteenth century. Courtesy Lightner Museum, St. Augustine, Florida.

Polychrome Egyptian Revival stool made by Alexander Roux, circa 1865. Dimensions: 23 3/4" high. Courtesy Metropolitan Museum of Art, New York. Purchase of the Edgar J. Kauffman, Jr., Foundation Gift, 1969.

Egyptian Revival rosewood cabinet. Center panel framed, two-tone bronze doré plaque of "Mitaros, Roi D'Egypt." Intricate inlay of classical and anthemion motifs. Made by Herter Brothers of New York, circa 1875. Courtesy Stingray Hornsby Antiques, Watertown, Connecticut.

French Empire/ Egyptian Revival style pedestal of carved, painted, and gilded mahogany with torch motif brass mounts. Late nineteenth century. Courtesy Lightner Museum, St. Augustine, Florida.

Berkey and Gay bed and dresser, 1876. Photo by George W. Davis, courtesy Pictorial Materials Collection, Grand Rapids Public Museum, Grand Rapids, Michigan.

Above left, left, and above:
Bed and dresser from an ebonized and carved Renaissance Revival bedroom set. Courtesy Neal Alford Auction Company.

Renaissance Revival bed, dresser, and wardrobe from a bedroom suite of carved and burled walnut, circa 1880. Photos courtesy Neal Alford Auction Company.

Single size sleigh bed by Alexander Roux.

Bureau and bed by T. Brooks and Company with the stenciled label of the maker. Courtesy Sotheby Parke Bernet Incorporated.

Renaissance youth bed. Collection of Steve and Barbara Mitchell, Kansas City, Missouri.

Renaissance Revival walnut child's bed/cradle with burl walnut trim. Collection of Richard and Eileen Dubrow.

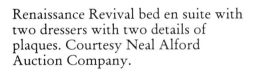

Renaissance Revival bed en suite with two dressers with two details of plaques. Courtesy Neal Alford Auction Company.

Rosewood dressing bureau labeled "Alexander Roux, New York." Courtesy EJ. Canton.

Rosewood bureau by Mitchell and Rammelsberg, Cincinnati, Ohio, circa 1865-1870. Dimensions: 99" high. Courtesy Newark Museum, Newark, New Jersey. Gift of Miss Grace Trusdell, 1926.

Sycamore bureau attributed to John Jelliff (1813-1893), made circa 1880-1890. Dimensions: 84" high. Photo by Stephen Germany, courtesy Newark Museum, Newark, New Jersey. Isabelle Ball Gifford Bequest, 1949.

Rosewood bureau by Alexander Roux, circa 1855. Dimensions: 96" high. Photo by Stephen Germany, courtesy of Newark Museum, Newark, New Jersey. Gift of Mrs. Lathrop E. Baldwin, 1975.

Walnut cabinet with burl walnut veneer and marquetry of lighter woods by Gotlieb Vollmer, Philadelphia, circa 1870. Dimensions: 41 3/4" high, 26 1/2" wide, 17 3/5" deep. Mark stenciled on underside of drawer reads: "G. VOLLMER,/CABINET & UPHOL-STERY/Warerooms & Manufactory,/1108 CHESTNUT STREET,and/1105 Sansom St./ Philadelphia." Courtesy Lyndhurst Corporation, New York.

Mahogany chest of drawers without hardware, signed "C.A. Baudoine, 333 Broadway, New York City." Courtesy Neal Alford Auction Company.

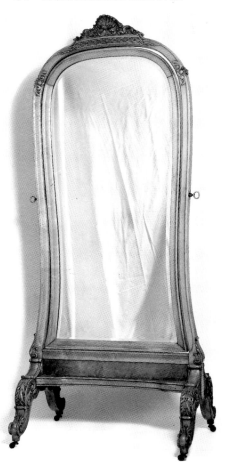

Fine Louis XV French style maple cheval mirror labeled "Horner, New York." Courtesy E. J. Canton.

Exceptional Renaissance Revival marble top dresser. Courtesy Neal Alford Auction Company.

Renaissance revival dresser en suite with bed and dresser. Courtesy Neal Alford Auction Company.

(See more bedrooms in color on page 73)

Walnut table with marble top. Photo by Phil Lewis, courtesy Kerr's Antiques, Marion, Indiana.

Walnut table with marble top. Collection of Yolanda and Jay Wright.

Ebonized and gilt stand. Collection of Yolanda and Jay Wright.

Inlaid table. Collection of Yolanda and Jay Wright.

Left: Inlaid tilt-top tea table. Courtesy Farm River Antiques.

Inlaid center table. Collection of Yolanda and Jay Wright.

Stand with inset marble. Collection of Yolanda and Jay Wright.

Center table by Wakefield Rattan Co., Boston, circa 1877. From the "Industrial Design" exhibit at the Grand Palais, Paris. Courtesy The Musee des Arts Decoratif.

Renaissance Revival table with ebonized and gilt trim and inlaid top. Detail view of inlaid top. Photos courtesy of Farm River Antiques.

Work, study, card, or office table by Clowes and Gares Manufacturing Co., circa 1877. From the "Industrial Design" exhibit at the Grand Palais, Paris. Courtesy The Musee des Arts Decoratif.

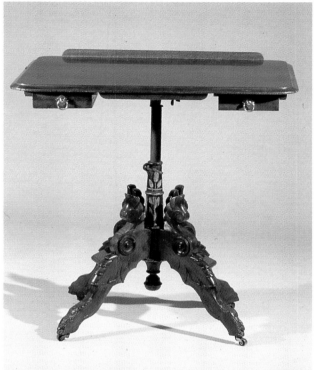

Rosewood center table with ebonized work, gilt incising, and brass beaded molding. Detail view of inlaid top. Courtesy Farm River Antiques.

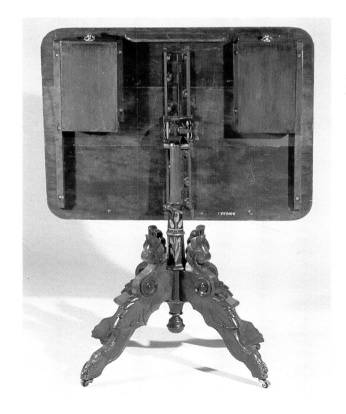

Three views of the only known model of a Moore Superior Grade desk. Walnut with various other woods inlaid. Courtesy Richard and Eileen Dubrow.

"Superior" grade Wooton desk shown closed, open with the writing board up, and open with the writing board down. Note the griffins on the top crest. Courtesy Richard and Eileen Dubrow.

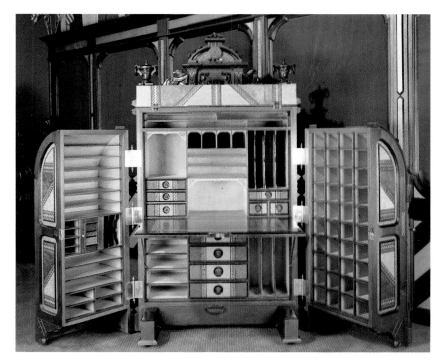

A rare "Ladies" Wooton patent secretary shown closed, open with the writing board up, and open with the writing board down. Courtesy Richard and Eileen Dubrow.

Superior Grade Wooton Patent Secretary made in 1875 in Indianapolis, Indiana. This was a custom order desk elaborately inlaid with covered cloisonné hardware and skillfully carved ornaments. Courtesy Richard and Eileen Dubrow.

Wiggers secretary bookcase with cabinet bottom shown with the writing surface opened and closed. Walnut with burl walnut trim and bird's-eye maple interior. Courtesy Richard and Eileen Dubrow.

Circausian walnut partners desk, heavily carved. Given by William R. Hearst to Marion Davis. Courtesy Richard and Eileen Dubrow.

Renaissance revival bedroom suite

Bedroom suite including bed,
dresser, and wash stand. Courtesy
Jay Anderson.

Bed of rosewood with marble top. Courtesy
Mr. J. Acunto.

Ornate bed by Mitchell and Rammelsberg. Courtesy
Joe Brosig.

Rosewood child's bed
by Alexander Roux.
Courtesy Farm River
Antiques.

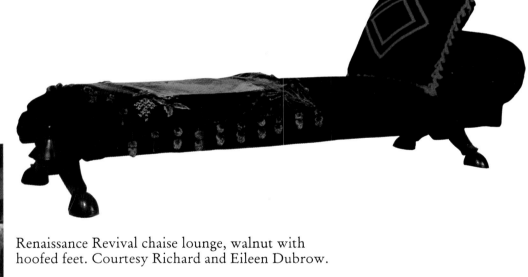

Renaissance Revival chaise lounge, walnut with
hoofed feet. Courtesy Richard and Eileen Dubrow.

Pair of Renaissance Revival twin beds, walnut
with burl walnut and incised gilt trim. Courtesy
Red Coach Antiques.

Renaissance Revival single night stand. Walnut with carved walnut pulls. Mid-West in origin. Courtesy Red Coach Antiques.

Wash stand, walnut with marble splash board and top. Photo by Phil Lewis, courtesy R. Beauchamp, Westfield, Indiana.

Wash stand, walnut with marble top. Photo by Phil Lewis.

Dresser, walnut with marble top. Dimensions: 95" high. Photo by Phil Lewis, courtesy Kerr's Antiques, Marion, Illinois.

Dresser, walnut with marble top. Photo by Phil Lewis.

Renaissance Revival walnut split pedestal table. Mid-West origin. Courtesy Red Coach Antiques.

Walnut sideboard with burl walnut trim. Photo by Phil Lewis, courtesy Victorian Galleries, South Bend, Indiana.

Rosewood cabinet-base sideboard stenciled "A. Roux." Collection of C.D. Hall.

Walnut and burl walnut, narrow marble top side board in the Renaissance Revival style. Courtesy Red Coach Antiques.

Walnut and burl walnut cabinet-based sideboard, probably mid-West in origin. Courtesy Mr. and Mrs. T. Merlo.

T. BROOKS & CO.
Furniture and Upholster
WAREHOUSE.
Nos. 127 & 129 FULTON ST.
COR. SANDS
BROOKLYN N Y

Superb walnut and burl walnut sideboard by T. Brooks and Company. Detail shows label from back of the piece. Photos courtesy Farm River Antiques.

Rosewood cabinet with ormolu mounts and beading by Alexander Roux. Labeled "A. Roux and Company." Courtesy Farm River Antiques.

Walnut and burl walnut renaissance revival cabinet with incised gilt trim and carving on the back splash, mid-West in origin. Courtesy Red Coach Antiques.

Renaissance Revival side cabinet, walnut with burl walnut and ebonized trim and gilt incising. Probably of mid-West origin. Courtesy Red Coach Antiques.

Rosewood cabinet with ebonized and gilt trim and bronze mounts. Courtesy Farm River Antiques.

Ebonized cabinet with bronze medallions and incised gilt lines.

Side cabinet with inlay panel, gilt and ebonized trim and porcelain plaque. Courtesy Farm River Antiques.

Renaissance Revival side cabinet. Rosewood with inlaid panels, brass trim and porcelain plaque. Probably of New York origin. Courtesy Red Coach Antiques.

Ebonized cabinet with gilt and brass trim and bronze plaques. Courtesy Farm River Antiques.

Table of black walnut attributed to Daniel Pabst, circa 1868. Dimensions: 28 1/2" high, 62 1/2" long, 50 1/2" wide. Courtesy Philadelphia Museum of Art, Philadelphia, Pennsylvania. Bequest of Arthur H. Lea. 31-1-63.

Renaissance Revival interior at the Fenno-Tudor House, Joy Street/corner Beacon. Courtesy of The Bostonian Society.

Renaissance Revival side board (one of a pair) labeled Alexandra Roux. Courtesy Margot Johnson.

Dining room table that opens for two 20" leaves. Labeled "Pillar Table/No. 252/Lenz's Pat. Sep. 7. 69/"Reissue May 3. 70/Pat. Mar. 17. 70/Lotz Pat. May 21.72." Collection of Steve and Barbara Mitchell, Kansas City, Missouri.

Rococo Revival carved rosewood étagère by J. and W. Meeks. Courtesy William Doyle Galleries, New York.

Red walnut sideboard by Daniel Pabst, Philadelphia, late nineteenth century. Dimensions: 68" x 25" x 87". Courtesy Philadelphia Museum of Art, Philadelphia, Pennsylvania. Bequest of Arthur H. Lea. F. 38-1-78.

Carved chestnut Renaissance Revival sideboard by Ignatius Lutz, Philadelphia, circa 1865. Dimensions: 94" high, 74" wide, 25" deep. Labeled Philadelphia furniture from this period is very rare and this is the only known signed example by this prominent Philadelphia firm, which is associated with some of the finest furniture of the 1850s and 1860s. Photo by Dick Goodbody, courtesy Kurland-Zabar, New York.

Sideboard circa 1860. Signed "A. Roux and Company, 479 Broadway and 43 and 46 Mercer Street, New York. French cabinetmaker and importers of fancy Buhl and Mosaic furniture. A. Roux, Joseph Cabus." Courtesy Christie's East, New York City.

Interior of miniature Renaissance Revival cabinet showing two small drawers. The interior is designed much like armoires of the period. Courtesy Red Coach Antiques.

Miniature Renaissance Revival one-door cabinet in walnut with burl walnut trim. Courtesy Red Coach Antiques.

Carved rosewood bookcase with stenciled label "E.W. Hutchings/Cabinet Warerooms/475/Broadway, NY." Courtesy Neal Alford Auction Company.

Walnut, maple, and poplar cabinet by Daniel Pabst of Philadelphia, circa 1865, 54" high, 44" wide, and 2 3/14" deep. Courtesy of permanent collection of High Museum of Art, Atlanta, Georgia. Gift from Virginia Carroll Crawford.

Rosewood, two-door bookcase stenciled T. Brooks.

Bookcase by G. Herter.
Collection of Mr. J. Bass.

Walnut Renaissance Revival parlor cabinet with gilt incised line decoration. Courtesy E. J. Canton.

Inlaid side cabinet of rosewood with gilt and ebonized trim. Courtesy Farm River Antiques.

Renaissance Revival Allen & Brothers cabinet ebonized with gilt trim. Front inlay panel is exceptionally fine. A very similar cabinet was displayed by this firm at the Philadelphia Centennial of 1876. *Gems of the Centennial Exhibition*, p.145.

Very fine, high style New York, rosewood, Neo-Grec parlor cabinet with incised line decoration, marquetry, ebonized trim, and fire gilt mounts. The inset panels show Ulysses and Penelope. Courtesy E. J. Canton.

Cabinet made by Pottier and Stymus for the Philadelphia Centennial exhibition (1876). Walnut with oak as a secondary wood, 138" high by 107" wide by 42" deep. Courtesy Post Road Gallery, Larchmont, New York.

Credenza by Gustave Herter (signed on back) with bronze plaque. Courtesy Post Road Gallery, Larchmont, New York.

Rosewood cabinet with rosewood veneer, marquetry of colored woods, brass plaque, and black marble top. Attributed to Herter Brothers of New York, circa 1865-1870, 47" high, 35" wide, and 21" deep. Courtesy Lyndhurst Corporation, New York.

Rosewood, ebonized cabinet with inlaid panels and porcelain plaques. Collection of Yolanda and Jay Wright.

Rosewood credenza attributed to Alexander Roux. Courtesy Post Road Gallery, Larchmont, New York.

Rosewood cabinet in Neo-Grec style, late nineteenth century, American. Courtesy Art Institute of Chicago, Chicago, Illinois.79.502.

Renaissance Revival rosewood cabinet by Alexander Roux, circa mid-1860s. American, Neo-Grec style. Dimensions: 53 5/8" high. Courtesy Metropolitan Museum of Art, New York, New York. Edgar J. Kauffman Charitable Foundation Fund, 1968, 68.100.1.

Very fine high style Louis XVI Revival rosewood three-door parlor cabinet with incised line decoration, marquetry panels and applied fire gilt mounts. The central ebonized door has a porcelain plaque flanked by maple marquetry panel doors. The mounts are marked P&S for Pottier & Stymus of New York. Courtesy E. J. Canton.

High style rosewood parlor cabinet with marquetry of various woods, ebonized trim, fire gilt mounts, and a central porcelain plaque. The piece is labeled by Thomas Godey of Baltimore. Courtesy High Museum, Atlanta.

Window cornice in the Renaissance Revival style of ebonized wood with incised gilt lines and draped silk. Collection of Yolanda and Jay Wright.

Renaissance Revival walnut and wicker log basket with needlework trim. Courtesy of Neal Alford Auction Company, Incorporated, New Orleans, Louisiana.

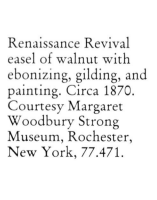

Renaissance Revival easel of walnut with ebonizing, gilding, and painting. Circa 1870. Courtesy Margaret Woodbury Strong Museum, Rochester, New York, 77.471.

Renaissance Revival globe stand with original Rand McNally globe. Collection of Steve and Barbara Mitchell, Kansas City, Missouri.

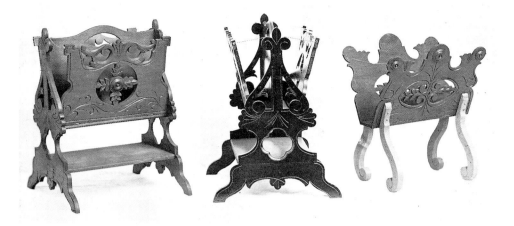

Magazine and folio racks with the trade card of E. M. Tibbetts Company, undertaker and furniture manufacturer of Dexter, Maine. Photo is marked "9 dollars per doz." on back. Courtesy Richard and Eileen Dubrow.

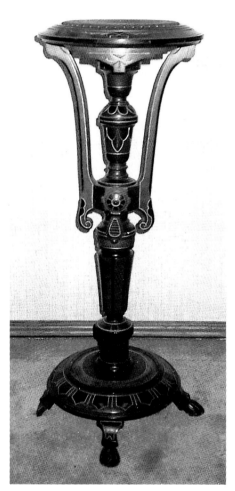

Plant stand. Collection of Steve and Barbara Mitchell, Kansas City, Missouri.

Music stand with inlaid, veneered, ebonized, carved, gilded, and incised wood, plus bronze mount, late nineteenth century. Courtesy Lightner Museum, St. Augustine, Florida. 86.1.134.

Pedestal of ebonized wood with 88 ormolu mounts and ceramic insets. Late nineteenth century. Courtesy Lightner Museum, St. Augustine, Florida.

Fern table of stick and ball style walnut with brass studs and bands, 32 1/4" high. Courtesy Norman Mizuno Fine Art and Antiques.

Black walnut étagère with tulip, poplar, and marble with mirror. Made by George W. Ware and Company, Boston, Massachusetts. Circa 1865. Courtesy Virginia Museum of Fine Arts, Richmond, Virginia. Mary Morton Parsons Fund. 78.31.

Hall mirror of Renaissance Revival style in walnut and burl walnut with gilt incising.

George Hunzinger considered himself a cabinet-maker, however he was more than that. Like John Henry Belter, George Hunzinger patented designs that improved upon the established cabinetmaking techniques. While he used standard Renaissance revival motifs like acorn finials, nail heads, incised lines and ebonizing, he also moved the angle of the front legs on some chairs so that they attached to the back upright as well as to side seat rail. He patented a coated wire mesh that was used on chair seats and backs. He also held patents on folding chairs and on a duplex spring for his platform rockers.

In all, Hunzinger held 20 patents. Almost all of Hunzinger's work carries a "patent brand" or paper label to protect his patent rights. Many chairs had cog-like turnings of almost modern machine-like appearance. All of his furniture was useful, innovative, and sometimes amusing.

Mahogany platform rocker by George Hunzinger, 42 1/2" high. Collection of Norman Mizuno Fine Art and Antiques.

Oak-stained mahogany "lollipop" rocker by George Hunzinger, 42" high. Collection of Norman Mizuno Fine Art and Antiques.

G. Hunzinger "coated wire mesh" day bed. Frame made of hard maple. Courtesy Richard and Eileen Dubrow.

Duplex spring "lollipop" platform rocker with continuous arms and back rail, patented by George Hunzinger. Collection of Richard and Eileen Dubrow.

Mahogany platform rocker with carved wolf head finials by George Hunzinger, 42 1/2" high. Collection of Norman Mizuno Fine Art and Antiques.

Mahogany platform rocker with carved lion head finials by George Hunzinger, 42" high. Collection of Norman Mizuno Fine Art and Antiques.

Oak platform rocker by George Hunzinger, 40" high. Collection of Norman Mizuno Fine Art and Antiques.

Maple Gothic-influenced rocker with geometric inlay and red incising by George Hunzinger, patent April 18, 1876. Dimensions: 36 1/2" high. Collection of Norman Mizuno Fine Art and Antiques.

Oak stained with walnut "lollipop" armchair by George Hunzinger, 34" high. Collection of Norman Mizuno Fine Art and Antiques.

Morris chair style as made by George Hunzinger. Courtesy Mr. and Mrs. Makoney.

Armchair of ebonized wood and gold incising by Hunzinger, patent March 3, 1869. Original silk upholstery with fringe, 37 1/2" high, 25 1/4" wide. Collection of Norman Mizuno Fine Arts and Antiques.

Rib cage armchair labeled George Hunzinger. Walnut with ebonized incizing, 39 1/2" high. Patented Mar 30, 1869. Courtesy E. J. Canton.

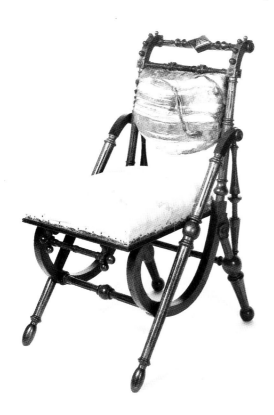

Side chair labeled George Hunzinger. Courtesy E. J. Canton.

Mahogany rocker with original leather seat by George Hunzinger, 42" high. Collection of Robert McMillen. Courtesy Norman Mizuno Fine Art and Antiques.

Walnut side chair mady by George Hunzinger, New York, circa 1869. Courtesy The Brooklyn Museum, Gift of Miss Isabel Shults by Exchange 1989.74.

Patent fancy armchair and side chair by George Hunzinger. Photo by Dennis Flynn.

Walnut Gothic-influenced side chair with ebonizing and incising by George Hunzinger, patent March 30, 1869, 33" high. Courtesy Norman Mizuno Fine Art and Antiques.

Folding rocking chair, walnut, brass, original upholstery. Made by George Hunzinger, circa 1870. Collection of The Brooklyn Museum, Department of Decorative Arts, George C. Brackett Fund, 1991.102a.

Folding chair by George Hunzinger, circa 1865.
Courtesy Neal Alford Auction Company.

Folding chair by George Hunzinger. Courtesy Certified Auctioneers and Appraisers.

Walnut side chair with ebonized incising by George
Hunzinge, 30" high by 46 3/4" long. Collection of
Norman Mizuno Fine Art and Antiques.

Walnut side chair with gold incising by George Hunzinger, patent March 30, 1869, 32 1/2" high. Collection of Norman Mizuno Fine Art and Antiques.

Walnut side chair by George Hunzinger, patent March 30, 1869, 33 1/2" high. Collection of Norman Mizuno Fine Art and Antiques.

Walnut side chair with ebonized incising by George Hunzinger, patent March 30, 1869, 29 1/4" high. Collection of Norman Mizuno Fine Art and Antiques.

Walnut side chair by George Hunzinger, patent March 30, 1869, 34 1/4" high. Collection of Norman Mizuno Fine Art and Antiques.

Walnut side chair by George Hunzinger, patent March 30, 1869, 34" high. Collection of Norman Mizuno Fine Art and Antiques.

Mahogany center table with metal mounted feet by George Hunzinger, 28" high. Collection of Norman Mizuno Fine Art and Antiques.

Oak tilt-top game table with walnut stain, checkerboard inlay, brass border, and felt top (reverse) by George Hunzinger, patent June 25, 1904, 28 1/2" high and 30 1/4" wide. Collection of Norman Mizuno Fine Art and Antiques.

Oak game table with storage drawers and four matching chairs by George Hunzinger, patent August 5, 1890, and May 24, 1892. Collection of Norman Mizuno Fine Art and Antiques.

Chapter 4
Aesthetic or Art Furniture Movement (1880-1900)

While the Renaissance Revival style was flourishing in its many and varied forms, a new aesthetic called "Art Furniture" (especially by designer Charles Eastlake) was coming into prominence. The firmer the Art Furniture's hold became, the more criticism was leveled at Renaissance styling. These critics and criticisms took furniture design in a new direction.

The British created the Art Furniture Movement from the Oriental inspired philosophies of Henry Cole and his circle, including furniture designers Bruce Talbert, William Goodwin, and Thomas Jekyll. It was promoted at the London Crystal Palace in 1851 and the New York Crystal Palace in 1853. At the Philadelphia Centennial of 1876, the Japanese exhibit became so popular and influential that it influenced even American furniture designers.

The change from heavy carving to incised lines had been a gradual one, a toning down from Rococo to Renaissance Revival. In the 1880s the jagged silhouette of furniture became smoother and more squared. Furniture was made with less exuberant decoration and fewer showy ornaments. This furniture generally has become known as the "Eastlake" style. Although the furniture was not designed by Charles Eastlake, it was based on his principles that craftsmanship should be visible and the finished piece devoid of veneer and ornament. The general idea of Eastlake furniture was to "improve taste in objects of modern manufacture" and to "encourage a discrimination between good and bad design." The style sometimes reflected modern Gothic types and sometimes combined Eastlake principles with Oriental and naturalistic motifs. The Oriental influence became strongest after the opening of the Japanese exhibit at the Philadelphia Centennial of 1876. Americans were fascinated by Japanese art forms and immediately copied the prominent but delicate designs that incorporated butterflies, fans, cherry blossoms, and chrysanthemums. It was not long before cabinetmakers included these elements as well as Japanese fretwork into their designs. Seating furniture continued to betray vestiges of Renaissance Revival styling, but the absence of heavy carving places it among Eastlake or Art Furniture.

In the factory-produced furniture of Grand Rapids, much of the subtlety and essence of the Eastlake principles were missing; however, the style was and continues to be known as Eastlake furniture.

The leaders in designs for Aesthetic or Art Movement furniture were Herter Brothers, Kimbel & Cabus, L. Marcotte, Herts Brothers, Pottier and Stymus, Roux and Company, and Sypher and Company.

Walnut sofa with carved heads. Part of a four-piece set. Probably of mid-West origin. Collection of Steve and Barbara Mitchell, Kansas City, Missouri.

Interior of the Oliver Ames House, Commonwealth and Massachusetts Avenues, Boston, showing an Aesthetic period styling. Courtesy The Bostonian Society.

Aesthetic style room interior in Rochester, New York.

Interior of Gross Mansion, showing a Moorish interior of the period. Courtesy Lightner Museum, St. Augustine, Florida.

Interior of Gross Mansion, showing a different view. Courtesy Lightner Museum, St. Augustine, Florida.

Aesthetic style tables and chair designed by Louis C. Tiffany. Courtesy Shelburne Museum Incorporated, Shelburne, Vermont.

Moorish/Turkish Revival corner chair with original upholstery. Courtesy E. J. Canton.

Ebonized walnut armchair with gold incising and carved dog head hand grips, 36 1/2" high. Collection of Norman Mizuno Fine Art and Antiques.

Walnut armchairs by Pottier and Stymus, 1875, 51 3/4" high. Courtesy Metropolitan Museum of Art, New York. Gift of Auguste Pottier, 1888.

Side chair, settee, and armchair by Herter Brothers. Courtesy Lockwood Mathews Mansion.

Fine ebonized curule-based armchair with incised line decoration and original upholstery, attributed to Herter Brothers. Courtesy E. J. Canton.

Pair of ladies' chairs with inlaid wood and gilt incised decorations.

Gilded maple side chair with mother-of-pearl inlay and embroidered silk upholstery by Herter Brothers, New York City, circa 1880, 34 1/2" x 18 3/4" x 18". Courtesy Museum of Art, Carnegie Institute. DuPuy Fund.

Fine ebonized Anglo-Japanese lady's writing desk with marquetry panel and incised bamboo decorated drawer panels. Courtesy E. J. Canton.

High style ebonized Anglo-Japanese parlor cabinet with finely lacquered front doors in the Oriental tradition. Courtesy E. J. Canton.

Ebonized Anglo-Japanese parlor cabinet with inset beveled mirrors in the pagoda-form top. Courtesy E. J. Canton.

Rockefeller family
bedroom with
furniture by
George Schastey.
Courtesy Museum
of the City of
New York.

Ebonized cherry cabinet with imported tile
and metal hardware by Kimbel and Cabus,
New York City, circa 1876-1880. Courtesy
Hudson River Museum, Yonkers, New
York. Gift of Susan D. Bliss.

China cabinet attributed
to Herter Brothers with
an elaborately inlaid
floral band. Courtesy
Neal Alford Auction
Company.

Ebonized cherry cabinet with gilt and
marquetry of lighter woods by Herter
Brothers, New York City, circa 1880.
Dimensions: 60" high, 33" wide, 16 1/4"
deep. Courtesy permanent collection of the
High Museum of Art, Atlanta, Georgia. Gift
from Virginia Carroll. Crawford Collection.

Bed, mirror, chest of drawers, wardrobe, and side chair with inlaid ebonized wood by Herter Brothers, circa 1876. Courtesy Philadelphia Museum of Art, Philadelphia, Pennsylvania. Given by Mrs. William T. Carter.

Bedroom suite by Herter Brothers, New York City, circa 1885, of ebonized wood with light wood floral marquetry and gilding. *Bed 76" high, 66 1/2" wide. *Dresser 33 3/4" high, 78 1/8" wide, 22 3/4" deep. *Bureau 30 13/16" high, 52" wide, 22" deep. *Wardrobe 83 5/8" high, 48 3/4" wide, 23 3/8" deep. *Night table 29 15/16" high, 17 1/8" wide and deep. *Pair of side chairs 34 3/16" high, 17 15/16" wide, 19 5/8" deep.

Sideboard of rosewood, ebonized cherry, maple, walnut, and satinwood by Herter Brothers, New York City, circa 1876-1884, 53" high, 71 3/4" wide, 16" deep. Courtesy Art Institute of Chicago. Gift of the Antiquarian Society through the Capital Campaign, 1986.

Cabinet branded Herter Brothers and with inlaid floral decoration. Courtesy Neal Alford Auction Company.

Breakfront cabinet signed "Herter," circa 1880, 91" high, 84" long. Courtesy Robert Bahssin, Post Road. Gallery, Larchmont, New York.

Carved ebonized-wood étagère stamped "Herts Brothers" on the back. Courtesy Post Road Gallery, Larchmont, New York.

Piano in ebonized cherry, bird's-eye maple, and gilt by Hallet, Davis and Company of Boston, Massachusetts, circa 1876. Dimensions: 95 inches high, 82 1/2" wide, 31 3/4" deep. Courtesy of the permanent collection of the High Museum of Art, Atlanta, Georgia, 1983. Virginia Carroll Crawford Collection.

Rosewood cabinet (circa 1880-1890) with rosewood veneer, American black walnut, mother-of-pearl, brass, copper, and pewter by Herts Brothers, 59 1/2" x 59 3/4" x 15 3/4". Permanent collection of the High Museum of Art, Atlanta, Georgia, 1983. Virginia Carroll Crawford Collection.

Branded Herter Bros. cabinet ebonized with inlaid decoration and gilt trim. Courtesy Richard and Eileen Dubrow.

Walnut cabinet with inlaid burl ash and inset glass panels made by Daniel Pabst, Philadelphia, circa 1875. Courtesy The Brooklyn Museum, Bequest of Marie Bernice Bitzer by Exchange 1990.9.

Gothic walnut sideboard by Kimbel and Cabus, circa 1876, 80" high, 70" wide, 26 1/4" deep. An almost identical sideboard appears in an illustration of the Kimbel and Cabus showrooms from 1877. Photo by Dick Goodbody, courtesy Kurland-Zabar, New York.

Aesthetic movement Walnut hall stand by Herter Brothers. Courtesy Sotheby Parke Bernet.

Aesthetic Movement presentation easel of ebonized cherry, elaborately carved griffin supports, and carved "Japonesque" floral panels by Herter Brothers, New York City, circa 1875. Courtesy Stingray Hornsby Antiques, Watertown, Connecticut.

Aesthetic Movement rosewood-inlaid bird's-eye maple fall-front desk stamped "Herter Brothers." Made for Edward H. Harriman, New York, 1879. Courtesy Margaret Caldwell.

(Art Furniture chapter continues on page 145.)

Walnut and gilt cornice stamped by Herter Brothers. The cornice is from Thurlow Lodge, the home of former California governor Milton Latham. Courtesy David Hill Asian Art.

Renaissance Revival canterbury. Courtesy Gene Canton.

Renaissance Revival pedestal. Courtesy Gene Canton.

Magazine/folio rack of walnut with ebonized sections and gilt trim. Photo by Phil Lewis, courtesy Kerr's Antiques, Marion, Illinois.

Garden chair made with strips of steel by Lalance and Grojean, New York, circa 1866. From the "Industrial Design" exhibit at the Grand Palais, Paris. Courtesy The Musee des Arts Decoratif.

Exceptional and nearly matching billiard table, c. 1880, and pool table, c. 1885, of rosewood and walnut with firefighting scenes of inlayed wood including steam pumpers pulled by two horses, an early fire mark, standing firemen with hoses and water shooting out, the American flag, and, dramatically, firemen coming out of a burning building.

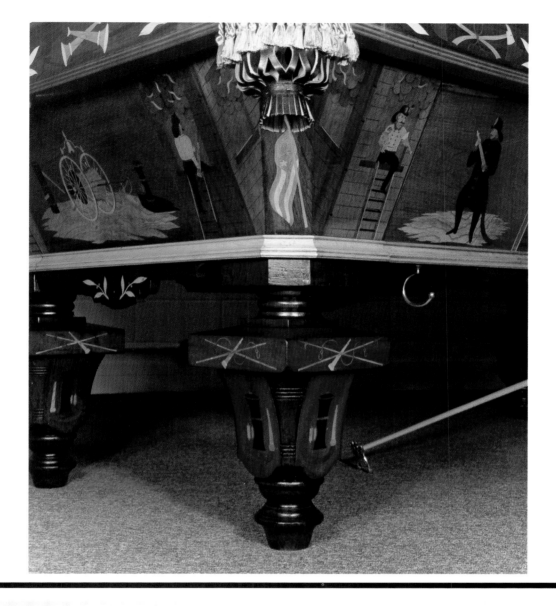

Duplex spring wire seat platform rocker patented by George Hunzinger (two views).

Arm chair with woven wire back and seat made by Georg Hunzinger, New York, circa 1875. From the "Industrial Design" exhibit at the Grand Palais, Paris. Courtesy The Musee des Arts Decoratif.

Duplex spring rocker patented by George Hunzinger.

Duplex spring "lollipop" platform rocker with high back, patented by George Hunzinger. Courtesy Mr. & Mrs. Manshel.

Hunzinger folding chair and continuous arm "lollipop" platform rocking chair.

Flip-top game table by George Hunzinger, patented June 26, 1894. Courtesy Peter Sena.

Ebonized chair with gilt trim by George Hunzinger. Courtesy Mrs. L. Aupperlee.

Aesthetic Movement couch attributed to Herter and Pottier. Gilt, figural arm supports, porcelain plaques, inlaid seat rail, and carved and inlaid decoration. Detail views of gilt arm support and porcelain plaque. Photos courtesy Mr. and Mrs. H. Gianotti.

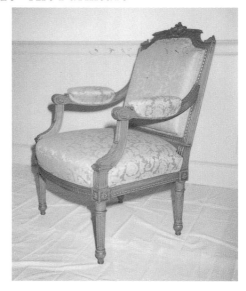

Arm chair attributed to Gotlieb Vollmer of Philadelphia. Courtesy R. Bloom, Norwalk, Connecticut.

Armchair similar to one published in *Art Decoration Applied to Furniture*, Chapter 15, "The Louis Quatorze," 1878, by Harriet Spofford. Courtesy R. Bloom, Norwalk, Connecticut.

Aesthetic movement ladies' chair with inlay on back splat and bronze medallion. Courtesy Richard and Eileen Dubrow.

Aesthetic movement bedroom set; dresser, armoire, and bed. Courtesy Mr. and Mrs. Robert Field.

Aesthetic movement bedroom set with dresser, bedstead, and wash stand, all with incised decoration. Courtesy Red Coach Antiques.

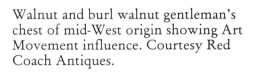

Walnut and burl walnut gentleman's chest of mid-West origin showing Art Movement influence. Courtesy Red Coach Antiques.

Ebonized cherry chest branded "Herter Bros.," elaborately inlaid in a floral design with Greek urns. It features three large drawers below a shallow top drawer that has interior fitting. Top is black marble with white and gold veining. Escutcheons, locks, and bail handles are brass. A variant, shown in the black and white photograph, has a white marble top. Courtesy Richard and Eileen Dubrow Antiques.

Bookcase with a Kimbel and Cabus label. Courtesy Post Road Gallery, Larchmont, New York

Piano case by Geo. Schastey for a Steinway model B (85 note) Serial #47245 piano built in 1882. This is probably the piano from the William Clark mansion shown in the book *Opulent Interiors of the Guilded Age*.

Étagère by Herter Bros. with their stamp on the back. Courtesy Ken Miller's Auction.

Arts and Crafts tile-top table by Gustav Stickley and tall bookshelves or magazine stand with a wood and copper sign inscribed with a motto "The ornaments of a home are the friends who frequent it" by Roycroft. Courtesy David Rago Arts & Crafts.

Group of Arts and Crafts furnishings. Left to right: Cutout armchair no. 575 with clip-corner arms and square cutouts on central back and side panels by Limbert, 41" x 27 1/2" x 22". --Lamp table no. 251 with overhanging octagonal top, long corbels, a flaring base, and trapezoidal cutouts by Limbert, 24 1/2" x 17" square. --Double oval library table no. 158 with overhanging oval top on flaring legs, oval tier and wide cross-stretchers with square cutouts by Limbert, 36" x 48" x 29". --Boudoir lamp in hammered copper and mica on a bulbous base by Dirk van Erp, 11" x 11". --Five-light chandelier no. 401 in handwrought copper embossed with hearts and bell-shaped amber glass shades by Gustav Stickley, 45" x 18" in diameter. Courtesy David Rago Arts and Crafts.

Trapezoidal single-door china cabinet no. 902 with pointed mullions, butterfly joints on chamfered doors, and inverted "V" apron by Gustav Stickley. This is one of Stickley's early pieces, 65" high, 37" wide, 15" deep. --Tall case clock with beveled overhanging top, incised copper face, arched apron, and clear glass door showing works by Leopold and J. George Stickley, 80" high, 25 1/2" wide, 15" deep. Courtesy David Rago Arts and Crafts.

Arts and Crafts style twin beds, pair of night stands, wash stand, and armoire with copper inlay attributed to the school of Koloman Moser. Courtesy Richard and Eileen Dubrow.

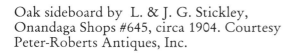

Oak sideboard by L. & J. G. Stickley, Onandaga Shops #645, circa 1904. Courtesy Peter-Roberts Antiques, Inc.

Sideboard #817 by Gustav Stickley, circa 1903. Courtesy Peter-Roberts Antiques, Inc.

Two door bookcase #717 by Gustav Stickley, circa 1910. Courtesy Peter-Roberts Antiques, Inc.

Single door china cabinet by Gustav Stickley, circa 1912. Courtesy Peter-Roberts Antiques, Inc.

Music cabinet #70 by Gustav Stickley, circa 1908.
Courtesy Peter-Roberts Antiques, Inc.

Rare Limbert library table and a nine-drawer chest by
Harvey Ellis, a designer in Stickley's Craftsman Workshops. Courtesy David Rago Arts & Crafts.

Oak desk with gallery, flush stretchers, and hammered
copper hardware by Gustav Stickley, Eastwood, New
York, circa 1907. Courtesy Cathers and Dembrosky
Gallery.

Oak drop front desk #550 by Gustav Stickley,
circa 1902. Courtesy Peter-Roberts Antiques, Inc.

Side chair with Eastlake influence and a peacock carved in the back. Courtesy George Wagner and Roberta Mayer.

Eastlake-influenced side chair with a crane design carved in the back. Courtesy George Wagner and Roberta Mayer.

G. Hoefstatter reclining chair. Courtesy Richard and Eileen Dubrow.

American bentwood back swivel office chair. Collection of Neal and Joyce Friedman.

Cane and black walnut swivel office chair, circa 1880. Courtesy North Wind Furnishings, Newton, Massachusetts.

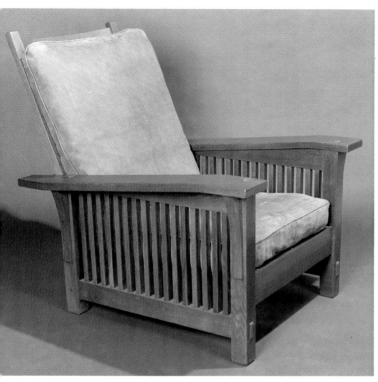

Drop arm Morris chair made by Gustav Stickley, circa 1907. Courtesy Peter-Roberts Antiques, Inc.

Oak hall chair by Elbert Hubbard's Roycroft Shops, circa 1900. Courtesy Cathers and Dembrosky Gallery.

Knockdown bed by Gustav Stickley, circa 1907. Courtesy Peter-Roberts Antiques, Inc.

Knock down settle by Gustav
Stickley, circa 1903. Courtesy
Peter-Roberts Antiques, Inc.

Settle by Gustav
Stickley, circa
1912. Courtesy
Peter-Roberts
Antiques, Inc.

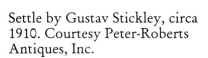

Settle by Gustav Stickley, circa
1910. Courtesy Peter-Roberts
Antiques, Inc.

Oak and tooled leather screen by Gustav Stickley, circa 1904. Courtesy Cathers and Dembrosky Gallery.

Gong of oak and hammered copper by Gustav Stickley, circa 1904. Courtesy Cathers and Dembrosky Gallery.

Oak and copper monumental candlesticks by Charles Rohlfs, Buffalo, New York, circa 1901. Courtesy Cathers and Dembrosky Gallery.

Arm chair and settee with
bentwood frames by Thonet.
Courtesy Tom Baier Assoc.
Auctioneers, Rick Black photog-
rapher.

Signed Thonet child's desk and
chair combination. Courtesy
Richard and Eileen Dubrow.

Bent plywood child's
chair labeled "Thonet
Child's Chair," circa
1945. Courtesy Richard
and Eileen Dubrow.

American oak bentwood settee. Collection of Neil
and Joyce Friedman.

American Art Nouveau uphol-stered couch. Courtesy Red Coach Antiques.

Art Nouveau-influenced golden oak lady's desk with original finish by Paine's Furniture, Boston, circa 1900. Courtesy North Wind Furnishings, Newton, Massachusetts.

Upholstered easy chair designed by Donald Deskey.

Club chair with wood arms. American, designer unknown, 1938.

Club chair with walnut cube designed by Edward Snohr, New York, 1929.

Blonde mahogany and leather arm chair designed and made by Modernage Furniture Co., circa 1933, New York.

Two seat sofa with a bent wood frame designed by Gilbert Rohde. Manufactured by Heywood Wakefield, Gardner, Massachusetts, 1931.

Side chair with wood frame, American, circa 1928.

Arm chair with wood frame
designed by Eugene Schoen,
New York, 1928.

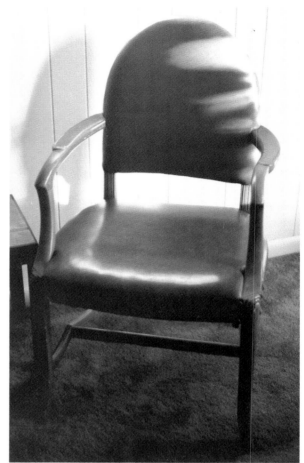

Custom made game table
and eight chairs. Designer
and maker unknown,
made in 1930.

Lounge chair with a tubular chrome frame designed by Gilbert Rohde. Manufactured by Troy Sunshade Co., Troy, Ohio, 1937.

A stool of tubular chrome designed by Gilbert Rohde circa 1936. Manufactured by Troy Sunshade Co., Troy, Ohio.

Lounge chair with tubular chrome frame, designer unknown. Manufactured by Royal Chrome Co., St. Louis, Missouri, 1938.

Side chair with aluminum frame designed by Warren McArthur, Rome, New York, 1935.

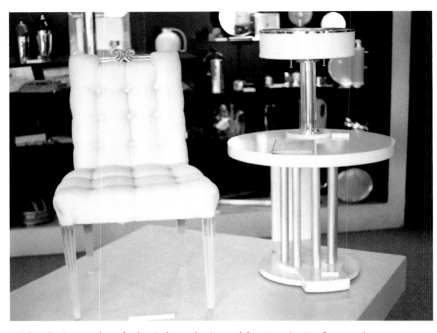

Lounge chair with tubular metal arms and legs and upholstered cushion seat and back designed by Kem Weber for Lloyd Manufacturing. Arm chair of tubular aluminum designed by Warren McArthur, c. 1935. Courtesy Don Treadway Gallery.

Side chair made of plexiglass designed by Lorin Jackson circa 1940. Manufactured by Grosfield House, New York. The table is made from aluminum and lacquered wood and designed by James Mont circa 1938. The manufacturer is unknown, New York.

Part of a line called Glassic, this plexiglass bench was designed by Lorin Jackson circa 1940. Manufactured by Grosfield House, New York.

Arm chair made of extruded aluminum designed by Warren McArthur. Manufactured by Warren McArthur, Rome, New York, circa 1936.

Vanity made of east Indian laurel glass and vinyl. Designed by Gilbert Rohde. Manufactured by Herman Miller Fur. Co, Zeeland, Michigan, 1938.

Walnut table designed and manufactured by Edward Snohr, New York, 1929.

Walnut table designed and manufactured by Edward Snohr, New York, 1929.

Rosewood, sharkskin, and pewter table designed and manufactured by Edward Snohr, 1929.

Vanity made of lacquered wood designed by Paul T. Frankl. Manufactured by Frankl Galleries, New York, 1929.

Blonde mahogany, brass and vinyl desk designed by Gilbert Rohde. Manufactured by Herman Miller Co., 1939.

Desk made of walnut burl and pewter. Designed and manufactured by Edward Snohr, New York, 1929.

Chest made of lacquered wood. Designed by Bet Wolk, manufacturer unknown, New York, 1928.

Chest bookcase of walnut and pewter designed and manufactured by Edward Snohr, New York, 1929.

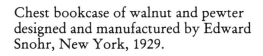

Console table of square tubular chrome and wood. Designer and maker unknown. Circa 1935.

A plexiglass table designed by Lorin Jackson. Manufactured by Grosfield House, New York, circa 1940.

Aluminum and walnut step table, designer and maker unknown. Circa 1935.

Coffee table of chrome and glass designed by Donald Deskey, 1935.

Coffee table of tubular copper and glass. Designed and made by Modernage Furniture Co., New York City, 1938.

Two ash stands of brass, chrome, and Bakelite designed by Walter Von Nessen circa 1932. Manufactured by Chase Copper and Brass Co., Waterbury, Connecticut.

Desk of walnut brass and vinyl designed by Gilbert Rohde, manufactured by Herman Miller Co., Zeeland, Michigan, 1939.

Chest and mirror made of plexiglass wood and mirror. Designed by Lorin Jackson, manufactured by Grosfield House, New York, 1940.

Desk and bookcase designed by Ralph Widdicomb, manufactured byWiddicomb, Grand Rapids, Michigan, 1932.

Desk by Kimbel and Cabus. Arabic inscriptions in panels. Courtesy Smithsonian Institution, Washington, D.C.

Fine ebonized Neo-Grec pedestal. Courtesy E. J. Canton.

Center table of ebonized walnut, gray marble top, incised decoration, and carved faces, 30" high and 32" wide. Collection of Norman Mizuno Fine Art and Antiques.

Aesthetic Movement ebonized cherry plant stands. The bases, punctuated by water and rock design and resting on stylized paw feet, are of great originality. Circa 1875. Courtesy Stingray Hornsby Antiques, Watertown, Connecticut.

Ebonized Anglo-Japanese fire screen with etched glass panel. Courtesy EJ. Canton.

Aesthetic Movement ebonized display cabinet with incised decoration and two inset tiles by W.B. Simpson and Sons, cabinet made by Kimbel and Cabus, New York City, circa 1876. Dimensions: 54 1/2" high, 41" wide, 11 1/4" deep. (A very similar cabinet with identical tiles is illustrated in the Kimbel and Cabus catalog at the Cooper-Hewitt Library.) Photo by Dick Goodbody, courtesy Kurland Zabar, New York.

Modern Gothic credenza of ebonized cherry with gilt incised decorations. Gold leaf floral and allegorical polychromed panels with elaborate hardware. Exhibited at the 1876 Centennial Exhibition. A companion piece is in the Victoria and Albert Museum, London, England. Courtesy Stingray Hornsby Antiques, Watertown, Connecticut.

Two étagères by Kimbel and Cabus. Courtesy
Smithsonian Institution, Washington, D.C.

Two étagères by Kimbel and Cabus. Courtesy
Smithsonian Institution, Washington, D.C.

Cast-iron fireplace mantle with inset reverse painted glass panels of the
Capitol at Washington, D.C., and Mount Vernon, home of George Washington. Courtesy E. J. Canton.

Chapter 5
Arts and Crafts Movement and the "Prairie School" Furniture (1890-1920)

The Arts and Crafts Movement was an approach to furniture that contained the more advanced aspects of Japanese styling even though it was severely geometric. The style repeated slats rhythmically, radiated the stretchers, and mullioned the case pieces. Like Renaissance Revival before it, Arts and Crafts furniture was not a single style but a sharing of a basic ideal of reform that had begun in England. Its object was to improve the quality of design and craftsmanship and to break down the barrier between fine and applied art.

The movement began in England to correct what designers felt was an imbalance brought on by specialization in furniture making and a decline in the quality of design and craftsmanship. Followers felt that mass production and labor specialization left the furniture designer with no firsthand knowledge of the methods of production and therefore left him unable to furnish designs appropriate for particular materials.

The aim of this reform movement was art, not design, but no reformer succeeded in producing a form as timeless or as modern as Thonet. He achieved the unity of technology and aesthetics that the other reformers only talked about.

"Prairie School" furniture was a derivative of the Arts and Crafts Movement, using built-in designs to define space and movable pieces to create screens. The movement was more interested in bringing hand and machine work together than in being strict to Morris or even the Greene Brother's ideas. Morris would have banned the machine. The Prairie School's most important tenet was that the building

and its furnishings were one. The school began in the office of architect Joseph Silsbee in the 1880s and grew up around Frank Lloyd Wright. The spirit of the movement was given form by architect Louis Sullivan. Its name derives from the low, horizontal houses these architects designed to conform with the flatness of the prairie. Many of these architects did not execute their own furniture designs but gave them to others for manufacture. Most architects of the Prairie School recommended Stickley furniture for secondary rooms.

It should be acknowledged that the Chicago fire of 1871 created opportunities for these architects to design houses and buildings and try out their ideas.

The Arts and Crafts Movement in Chicago and particularly in Cincinnati also greatly encouraged women furniture makers, carvers, and decorators. Women were trained by Isaac Scott at the Chicago Society of Decorative Arts and encouraged to support themselves. They were given free classes in wood carving, painting, and needlework. They also maintained a salesroom to sell the pieces they made.

Dining room in the Bolton House, 1907, designed by Greene and Greene. Courtesy Documents Collection, College of Environmental Design, University of California, Berkeley.

Breakfast room in the George H. Barker House designed by Greene and Greene, Pasadena, California. Courtesy Documents Collection, College of Environmental Design, University of California, Berkeley.

Arts and Crafts style clothes press by Isaac Scott. Courtesy Chicago Architecture Foundation.

Small bookcase by Isaac Scott. Courtesy Chicago Architecture Foundation.

Large bookcase by Isaac Scott. Courtesy Chicago Architecture Foundation.

Walnut cabinet with inset painted (polychrome and gilt) panels of medieval musicians and inscription "God Bless You Merry Gentlemen Let Nothing You Dismay." Courtesy E. J. Canton.

Double-door bookcase no. 703 designed by Harvey Ellis. Courtesy David Rago Arts and Crafts.

Smoking stand by Gustav Stickley, circa 1904.

Oak bride's chest with iron hardware by Gustav Stickley, circa 1902. Courtesy Cathers and Dembrosky Gallery.

Chest of drawers no. 602 by Gustav Stickley with two over four drawers, inverted "V" backsplash, chamfered sides, and through tenons, 53 1/2" high, 40" wide, 22" deep. Courtesy David Rago Arts and Crafts.

Cincinnati Art Furniture

In Cincinnati the reform movement was set in motion by Henry Lindley Fry, his son William Henry Fry, and Benjamin Pitman. Henry and William Fry were commissioned to decorate "Rookwood," the home of Joseph Longworth. This commission gave social acceptability to the movement. The Frys also worked on the house of Longworth's daughter, Maria Longworth Nichols. Cincinnati furniture leaned heavily on Gothic Revival and followed many of the principles of Pugin and Scott, even to the point of leaving surfaces unvarnished. They differed from Eastlake's principles in that they did not restrain the use of ornament. The Frys championed wood carving as women's art and advocated women as beautifiers. Along with Benjamin Pitman, the Frys founded the School of Design of the University of Cincinnati in 1872. The Cincinnati Art Furniture Movement lasted until the death of William Fry in 1929.

Chest of drawers, probably American, black walnut, made by Agnes Pitman of Cincinnati, 1876. Courtesy of Cincinnati Art Museum, Gift of Miss Melrose Pitman 1970.164.

Walnut chest of drawers designed and carved by Benn Pitman of Cincinnati, 1884, 45" high, 39 3/4" wide, 18 1/2" deep. Courtesy of Cincinnati Art Museum, Gift of Miss Melrose Pitman 1970.163.

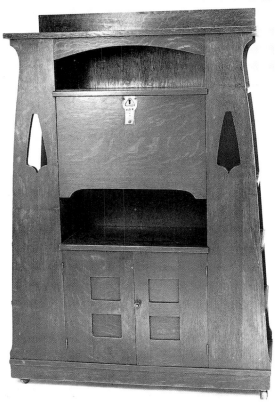

Writing desk, letter case, and lamp frame from the Gamble House living room, designed by Greene and Greene. Courtesy Documents Collection, College of Environmental Design, University of California, Berkeley.

Rare and early drop-front desk in original dark finish by Limbert, circa 1905. Courtesy David Rago Arts and Crafts.

 Desk and chairs in the living room of the Cordelia A. Culbertson House, Pasadena, California, circa 1911, designed by Greene and Greene. Courtesy Documents Collection, College of Environmental Design, University of California, Berkeley.

☞

Library table by Isaac Scott. Courtesy Chicago Architecture Foundation.

Desk by Isaac Scott. Courtesy Chicago Architecture Foundation.

Library table designed and carved by Benn Pitman of Cincinnati, circa 1878, 36" high, 23 1/4" wide, 16 1/2" deep. Courtesy of Cincinnati Art Museum, Gift of Miss Melrose Pitman.

Table made by William Henry Fry of Cincinnati, 29" high, 32" diamter. Courtesy of Cincinnati Art Museum, Gift of Mrs. Ida McGowen 1926.56.

Table made by William Henry Fry of Cincinnati, late 19th century, 28" high, 41 7/8" wide, 27" deep. Courtesy Cincinnati Art Museum, Gift of Mrs. James Morgan Hutton 1964.183.

Taboret (a small table or stand) with clipped corners, one octagonal Grueby Co. ceramic tile in the center, cross-stretchers, and legs mortised through the top, designed by Gustav Stickley, 22" x 17" x 17". Courtesy David Rago Arts and Crafts.

Flared leg plant stand or drink stand by Leopold and J. George Stickley.

Oak, hexagonal, leather-covered library table with brass tacks by Gustav Stickley, Eastwood, New York, circa 1903. Courtesy Cathers and Dembrosky Gallery.

Oak rocking chair by Charles Rohlfs, 1901. Courtesy Carnegie Museum of Art, Pittsburgh, Pennsylvania, DuPuy Fund.

White oak armchair inlaid with copper, pewter, and wood by Harvey Ellis, circa 1903. Courtesy Carnegie Museum of Art, Pittsburgh, Pennsylvania, Decorative Arts Purchase Fund: Gift of Mr. and Mrs. Aleon Deitch by exchange.

Oak side chair with copper faces of men and brass studs, 31 3/4" high. Collection of Norman Mizuno Fine Art and Antiques.

Drop-arm Morris chair no. 369 with five slats under each arm, long corbels, and through tenons, 40" x 33" x 38". Taboret (small table or stand) no. 601 with overhanging top, arched cross-stretchers, fully pegged, 16" x 14", both by Gustav Stickley. Courtesy David Rago Arts and Crafts.

High-back armchair with arched apron and inlaid with stylized flower design in copper, nickel, and wood by Harvey Ellis, circa 1904, 44" high, 25" wide, 21 1/2" deep.

Bow-arm Morris chair by Leopold and J. George Stickley. Footstool and spindled library table by Gustav Stickley. Courtesy David Rago Arts and Crafts.

Tall-back spindled rocker with black finish by Gustav Stickley.

High-backed spindled armchair no. 386 by Gustav Stickley, octagonal taboret (small table or stand) no. 515 and spindled side chair by Leopold and J. George Stickley. Courtesy David Rago Arts and Crafts.

Cube sofa with spindles and chip-corner lamp table by Gustav Stickley. Courtesy David Rago Arts and Crafts.

Even-armed, spindled settee with four higher corner posts, fully pegged, spring seat, by Gustav Stickley, circa 1905. Courtesy David Rago Arts and Crafts.

Important mantel clock in oak case by L. and J.-G. Stickley. Courtesy David Rago Arts & Crafts.

Child's crib by Isaac Scott. Courtesy Chicago Architecture Foundation.

Umbrella stand, waste basket, and standing mirror by Gustav Stickley.

Mantel made by Henry L. Fry and William H. Fry of Cincinnati, circa 1851, 50" high, 60 1/4" wide, 8 1/2" deep. Courtesy of Cincinnati Art Museum 1977.140.

Coat and umbrella costumer or hall stand by Limbert.

Michael Thonet

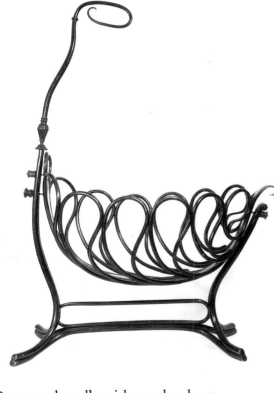

Bent mahoganized beechwood easel made by Michael Thonet, circa 1870. Courtesy Brooklyn Museum, Brooklyn, New York. Caroline A.L. Pratt Fund.

Bentwood cradle with overhead net hook labeled Thonet. Courtesy E. J. Canton.

Bent mahoganized beechwood stool with cane seat made by Michael Thonet, circa 1850. Courtesy Brooklyn Museum, Brooklyn, New York. Caroline A.L. Pratt Fund.

Rocking chair of bent, mahoganized beechwood upholstered in black leather. Made by Michael Thonet, circa 1860. Courtesy Brooklyn Museum, Brooklyn, New York. Caroline A.L. Pratt Fund.

Bentwood chaise lounge by Michael Thonet, circa 1870. Courtesy Brooklyn Museum, Brooklyn, New York. The Woodward Memorial Funds.

Beechwood child's armchair designed by Michael Thonet and made by Thonet Brothers, Moravia, circa 1875. Courtesy Brooklyn Museum, Brooklyn, New York. Gift of Barry Harwood.

Settee for dolls made of steam-bent beechwood. Probably Thonet Brothers, mid-1880s. Courtesy Brooklyn Museum, Brooklyn, New York. Anonymous gift.

(See more Prairie school furniture in color on page 124)

Chapter 6

Art Nouveau Furniture (1890-1910)

With the return of prosperity in the late 1890s, there was a renewed interest in more organic design. Art Nouveau emerged and was promoted as the "New Art" at the Paris Exposition of 1900. It had deep Neo-Rococo origins emphasizing free-flowing, asymmetrical, sinuous form and line. The furniture represented a break from academic tradition and an attempt to create a new style that had forms derivative of nature rather than historical precedent. Art Nouveau design philosophy demanded a room to be planned in its entirety; there could be no eclecticism, no pieces from other periods or times.

American craftsmen made significant contributions to almost every furniture movement with the exception of Art Nouveau; for it was not popular in America. It was not adaptable to the furniture problem of its time: large-scale machine production. Also, not enough working Americans were able to see the Paris furniture that first promoted this form. In the 1920s Art Nouveau furniture was reassessed, the superfluous decoration was abandoned, "form follows function" was a popular idea, and Art Deco emerged.

Rosewood arm chair inlaid with mother of pearl and brass, probaby from New York, circa 1885. A matching chair is in the Crawford Collection at the High Museum, Atlanta. Courtesy Stingray Hornsby Antiques.

Three-piece parlor set by Karpin Brothers. Courtesy Cincinnati Art Galleries, Cincinnati, Ohio.

Mahogany corner chair, circa 1900-1915, New York City area, 30 1/4". Courtesy Art Institute of Chicago, Chicago, Illinois. Dr. and Mrs. Edwin DeCosta Fund, 1979.

Oak side chair by Charles Rohlfs, circa 1898. Courtes6 The Art Museum, Princeton University.

American Art Nouveau mahogany side chair, circa 1895. Courtesy Neal Alford Auction Company.

Rare Art Nouveau Lady's desk of mahogany, mahogany veneers, and American sycamore, circa 1900-1910. The interior drawers are finished. Marks include a "9" impressed on the back of both upper drawers and a "7" upside down impressed on back of both lower drawers, 37 3/4" high, 29" wide, 22" deep. Courtesy Brooklyn Museum, Brooklyn, New York. H. Randolph Lever Fund.

Mahogany curio cabinet with glass and velvet, American, circa 1910, 58 3/4" high, 36" wide, 16" deep. Courtesy Metropolitan Museum of Art, New York. Edgar J. Kaufmann Charitable Foundation Fund, 1968.

(See more art nouveau in color on page 133)

Art Nouveau mahogany card table by Tobey Company of Chicago, circa 1910. Courtesy the Chrysler Museum, Norfolk Virginia. Gift of Walter P. Chrysler, Jr.

Art Deco furniture developed in the 1920s but was conceived much earlier, between 1905 and 1910, as an attempt to unite art and industry once again and end the excesses of Art Nouveau. It originally was thought of as "Art Moderne" and derived its name from the Paris Exhibition of 1925 (*L'Esposition International des Arts Decoratif et Industriels Modernes*).

Art Deco furniture emphasized geometric shapes and patterns. The seats of chairs and sofas were lowered and their backs were heightened. This change of proportion required low tables and therefore gave birth to the cocktail table. The woods favored for the furniture were exotic, and trim was of ivory and shagreen (an imitation sharkskin). Mother of pearl was used extensively.

The origins of Art Deco were in the geometric aspects of Egyptian and Mayan art, Cubism, Fauvism, and Expressionism. The opening of Egyptian King Tutankhamen's tomb in 1922 had a vast impact on design and designers worldwide. Art Deco was the first style to be totally and successfully mass produced. It harkened to the past only in its expression of luxury and in its cost. The period was luxury-loving as exemplified by its jewelry and its use of exotic woods in furniture.

Art Deco furniture breaks down to three styles:

Zigzag — geometric with stylized ornaments of zigzags, angular patterns, and abstract plant and animal motifs.

Streamlined Modern — futuristic with rounded corners and horizontal bands known as "speed stripes."

Classical Modern — simplified and monumental modernistic neoclassicism with an austere form of geometric and stylized relief sculpture.

Frank Lloyd Wright is the link in America between the Arts and Crafts and the Deco styles in his promotion of machine production of furniture and his Mayan and streamlined style houses. None-the-less, Art Deco furniture was strongly influenced by Art Nouveau with the sinuous lines of Nouveau being replaced by regular abstract lines.

Other leaders of Modernist furniture (American Art Deco) were Donald Deskey, Paul Theodore Frankl, Eugene Schoen, Karl Emanuel Martin (Kem) Weber, Gilbert Rohde, Ilonka Karasz, Joseph Urban, Jules Bouy, and Wolfgang Hoffman.

Mohair overstuffed armchair. Courtesy Savoia's Auctions Incorporated.

Kem Weber sofa for Lloyd loom. Courtesy Savoia's Auctions Incorporated.

Armchair with blond veneered legs and black plastic upholstery by Samuel Marx, 1944, manufactured by William Quigley Company, 30 3/8" high, 23 1/8" wide. Courtesy Art Institute of Chicago, Chicago, Illinois. Gift of Leigh B. Block, 1981.

Side chair with veneered legs and black plastic upholstery by Samuel Marx, American, circa 1944, manufactured by William Quigley Company. Dimensions: 23 1/2" high; seat is 18". Courtesy Art Institute of Chicago, Chicago, Illinois. Gift of Leigh B. Block, 1981.

Armchair of unidentified wood and leather by Kem Weber, 1928-1929. Dimensions: 40 1/2" x 21 1/2" x 20". Courtesy Art Institute of Chicago, Chicago, Illinois. Mr. and Mrs. Manfred Steinfeld, 1985.

Steel armchair by Donald Deskey, circa 1938, manufactured by Royal Chrome Company, 33 7/8" x 21" x 18". Courtesy Art Institute of Chicago, Chicago, Illinois. Gift of Mrs. Florence Schoenborn, 1970.

Serving table of unidentified wood and mahogany by Kim Weber, 1928-1929, 30" high, 16 1/2" wide, 36" in diameter. Courtesy Art Institute of Chicago, Chicago, Illinois. Mr. and Mrs. Manfred Steinfeld Fund, 1985.

Birch veneer side table by Samuel Marx, 1944, manufactured by William Quigley Company, 32 3/4" high, 66 1/8" wide, 24 1/2" in diameter. Courtesy Art Institute of Chicago, Gift of Leigh B. Block, 1981.

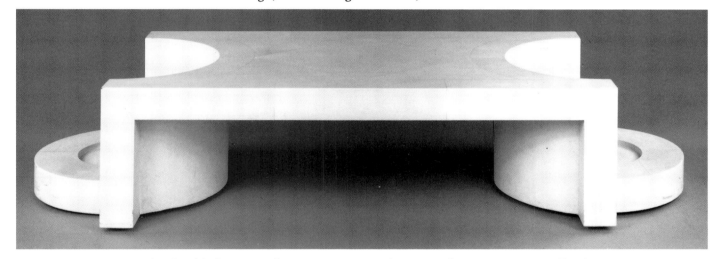

Cocktail table by Samuel Marx, 1944, 11 3/8" x 53 3/4". Courtesy Art Institute of Chicago, Gift of Mr. and Mrs. Alvin R. Whitehead, Jr., in honor of Leigh B. Block, 1985.

Cocktail table of parchment over birch frame by Samuel Marx, circa 1944, manufactured by William Quigley Company, 12 1/8" x 140" x 24". Courtesy Art Institute of Chicago, Gift of Leigh B. Block, 1981.

Birch with veneer end table by Samuel Marx, 1944, manufactured by William Quigley Company, 25 1/4" high, 17 15/16" wide. Courtesy Art Institute of Chicago, Gift of Leigh B. Block, 1981.

End table with white veneer finish and lucite by Samuel Marx, 1944, manufactured by William Quigley Company, 23 9/16" high, 14 1/16" diameter. Courtesy Art Institute of Chicago, Gift of Leigh B. Block, 1981.

Table made of aluminum and Bakelite designed by Walter Von Nessen. Manufactured by Newsen Studios, New York, 1929.

Cabinet by Eugene Schoen, 1928, 45" x 39 3/8" x 18 3/4". Courtesy Art Institute of Chicago, Restricted gift of Fern and Manfred Steinfeld, through prior bequest of Mr. Chester D. Tripp, 1988.

Pair if California redwood Skyscraper bookcases designed by Paul Theodore Frankl, 1920s, 90" high, 35 1/2" wide. Courtesy Cincinnati Art Museum, Gift of the Estate of Mrs. James M. Hutton II.

Walnut-veneered Art Deco dresser from a matched bedroom suite, probably designed by Ely Jacques Kahn, manufactured by Snohr of New York City, retailed by Mandel Brothers of Chicago, circa 1928. The low chest of this suite is identical to one pictured in a Kahn apartment in New York City on the cover of the 1929 *Furniture Buyer and Decorator*. The grid pattern and the fluting are typical of Kahn's furniture designs. The set is very progressive for American designs of this period. Photo by Dick Goodbody, courtesy Kurland-Zabar, New York.

(See more art deco in color on page 134)

Chapter 8:
Modern Furniture (1920–1960)

Furniture of the period 1920 to 1960 was concerned with form, whether machine-made or man-made, in wood, plastic, or metal. It emphasized sculpture, but always in an abstract way. The furniture never was ornamented, symbolic, or copied from nature. The designers attempted to simplify the manufacturing process by using a minimum of material, but tried to achieve comfort and convenience in the designs. While nineteenth century styles concealed the structure, twentieth century furniture emphasized it. The furnishings were lightweight, light in color, smooth of surface, had fewer joints, and an impersonal form. Many people have believed that Eames and Saarinen used plywood veneers in a revolutionary way (making multi-dimensional curves surpassing Alvar Aalto's early bent plywood). Actually, their designs continued J.H. Belter's ideas

as worked out in his spoon back chair, Samuel Crag's "elastic chair" of 1808, and Robert Adams's three-ply mahogany dining room chairs made for Osterley Park in 1773.

To this day, the furniture industry is conservative in its technology except for the few people involved with designs in plastics and metal. The industry as a whole has retained traditional materials and techniques and most often traditional forms as well. The "golden years" for modern furniture design in America were 1947 to 1960, an era of postwar prosperity, innovative design, and advancing technology. Partial responsibility for the decline of modern furniture after 1960 is the fact that the designers could not patent their creations, and therefore they were copied endlessly and often quite badly.

Gilded and silvered wood banquettes with crossed horn form legs by Victor Proetz, Saint Louis. Courtesy Selkirk Galleries, Saint Louis, Missouri.

Bedroom suite designed by Brower for the Sleigh Furniture Company circa 1880. Courtesy Grand Rapids Public Museum.

Alvar Alto upholstered bent birch chair and table. Courtesy Savoia's Auction.

Lacquered paper, veneered wood suite with dark blue/green molding including a cylindrical commode with white marble top, sleigh bed, round center table with gilt banding, dresser and mirror with giltwood sunburst in mirror and grey figured white marble. All pieces by Victor Proetz (1897-1966), Saint Louis. Courtesy Selkirk Galleries, Saint Louis, Missouri.

Round center table with white marble top inlaid with a central roundel of lapis lazuli and two bands of black marble by Victor Proetz, Saint Louis. Courtesy Selkirk Galleries, Saint Louis, Missouri.

Amboyna wood demi-lune commode with white marble top by Victor Proetz, Saint Louis. Courtesy Selkirk Galleries, Saint Louis, Missouri.

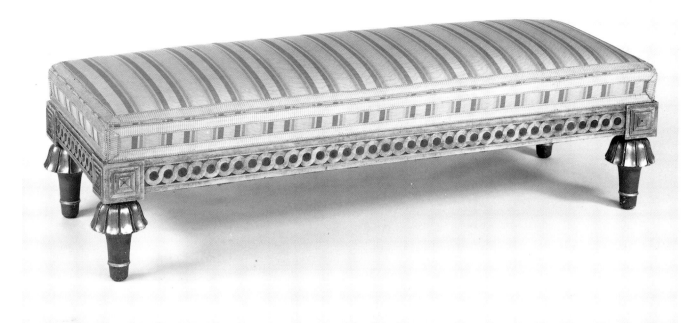

Carved and painted gilt wood window bench by Victor Proetz, Saint Louis. Courtesy Selkirk Galleries, Saint Louis, Missouri.

Long console table with pewter top (scored with a linear geometric motif), by Victor Proetz, Saint Louis. Courtesy Selkirk Galleries, Saint Louis, Missouri.

Pair of lamp tables with round, marquetry inlaid burlwood tops by Victor Proetz, Saint Louis. Courtesy Selkirk Galleries, Saint Louis, Missouri.

Black bookcase with stepped sides and red lining designed by Paul Theodor Frankl, 27 1/2" high, 78" wide. Courtesy Cincinnati Art Museum, Gift of the Estate of Mrs. James M. Hutton II, 1969.409.

Swan chair designed by Arne Jacobsen, 1950s. Courtesy Savoia's Auction Incorporated.

Molded plywood chair, the first version of which won for its designers Charles Eames and Eero Saarinen a first place in the Museum of Modern Art's "Organic Design in Home Furnishings" competition in 1940. Charles and Ray Eames revised the design for mass production, and Herman Miller began distributing the chair in 1946. The design combines wood and metal in a technologically innovative and aesthetically honest way that reveals mechanical connections, instead of disguising them. The rubber shock mounts that connect the bent chrome-plated steel rods to the walnut-veneer-faced plywood are plainly visible and as integral to the design as the elegant shapes of the seat, back, and frame.

Time magazine nicknamed it the "potato chip," and artist Saul Steinberg once drew it for the *New Yorker*—with an antimacassar draped over its backrest, implying how comfortable it is. Courtesy Herman Miller Archives, Zeeland, Michigan.

Arm chair designed by Peter Danko, Alexandria, Virginia, circa 1976. From the "Industrial Design" exhibit at the Grand Palais, Paris. Courtesy The Musee des Arts Decoratif.

Swivel chair designed by Raymond Loewy. Courtesy Savoia's Auction Incorporated.

Chair made with an inflated rubber tube covered with fabric, designed by Davis J. Pratt, circa 1948. From the "Industrial Design" exhibit at the Grand Palais, Paris. Courtesy The Musee des Arts Decoratif.

"Womb chair" designed by Eero Sarinen and manufactured by Knoll International, circa 1948. From the "Industrial Design" exhibit at the Grand Palais, Paris. Courtesy The Musee des Arts Decoratif.

LAR armchair made of molded plastic, designed by Charles Eames, circa 1949. From the "Industrial Design" exhibit at the Grand Palais, Paris. Courtesy The Musee des Arts Decoratif.

"Barwa" chair with tubular aluminum frame designed by Edgar Bartolucci and Jack Waldheim, circa 1947. From the "Industrial Design" exhibit at the Grand Palais, Paris. Courtesy The Musee des Arts Decoratif.

Assymetrical chest of drawers and night stand designed by Donald Deskey for Widdicomb Furniture Co., 1930. Courtesy Don Treadway Gallery.

Cone chair designed by Vernor Panton.
    Cabinet designed by George Nelson for Herman Miller. Courtesy Don Treadway Gallery.

Loveseat designed by Bruno Mathsson of molded birch laminate and natural fiber webbing.--Floor lamp designed by Iamu Noguchi for Akaiu.
Decorative table in the style of Piero Fornasetti.--Grasshopper chair designed by Eero Saarinen for Knoll. Courtesy Don Treadway Gallery

Stool, table and chair designed by Frank Gehry for Easy Edges, Inc. Courtesy Don Treadway Gallery.

Price Tower armchair and a skylight from the Avery Coonley House, both designed by Frank Lloyd Wright. Courtesy Don Treadway Gallery.

Molded fiberglass Shell arm chair designed by Charles and Ray Eames, 1948. This design was one of the first consumer products to make use of the high-performance materials developed by the aircraft industry during World War II. Made of fiberglass-reinforced polyester, it was a prizewinner in the Museum of Modern Art's "Low-Cost Furniture" competition in 1948. It was available on a variety of bases and in upholstered models. Courtesy Herman Miller Archives, Zeeland, Michigan.

Side chair designed by Donald Knorr for Kroll Assiciates.--Storage unit designed by Charles Eames for Herman Miller. Courtesy Don Treadway Gallery.

Lounge and ottoman designed by Charles Eames and originally built in 1956 as an evolution of 1940s experiments with molded plywood; it was not intended to be mass produced. But thirty years after Herman Miller began producing it, more than 100,000 were manufactured, and the Eames lounge has long been recognized as a symbol of quality. The chair is constructed of molded rosewood with leather cushions and a polished aluminum swivel base. Tufted leather cushions fit in the molded plywood forms. The base and back supports are attached with neoprene shock mounts that give the chair flexibility and resiliency. The chair won the Triennale Prize in Milan in 1957 and one is in the permanent collection of New York's Museum of Modern Art. Courtesy Herman Miller, Incorporated, Zeeland, Michigan.

Lounge chair and ottoman designed by Charles Eames and manufactured by Herman Miller, Inc. From the "Industrial Design" exhibit at the Grand Palais, Paris. Courtesy The Musee des Arts Decoratif.

Laminated and molded wood items designed by Charles and Ray Eames: dining chair made by Evans Product, folding screen mady by Herman Miller, dining table manufactured by Evans, and The Toy manufactured by Tigrett Toys. Courtesy Don Treadway Gallery.

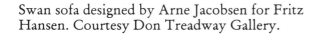

Swan sofa designed by Arne Jacobsen for Fritz Hansen. Courtesy Don Treadway Gallery.

Barcelona chair designed by Ludwig Mies van der Rohe for Kroll Internatioal. Courtesy Don Treadway.

Side chair with tubular steel frame and fiberglass back and seat. From the "Industrial Design" exhibit at the Grand Palais, Paris. Courtesy The Musee des Arts Decoratif.

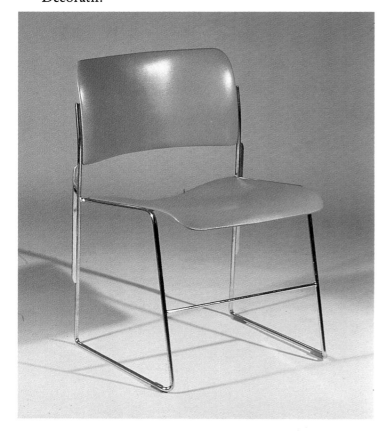

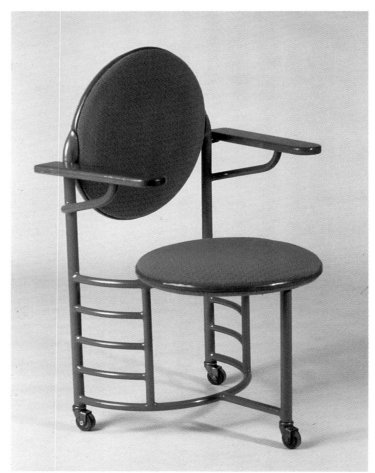

Arm chair designed by Frank Lloyd Wright for the S. C. Johnson Administration Building, circa 1936. From the "Industrial Design" exhibit at the Grand Palais, Paris. Courtesy The Musee des Arts Decoratif.

Molded fiberglass ganging/stacking Shell side chairs designed by Charles and Ray Eames, introduce in 1952. One of the bases allows the chairs to be stacked or hooked side by side into rows. This shell side chair won the first National Industrial Designers Institute Award medal in 1951, and it has been produced continuously by Herman Miller ever since.

"Diamond chair" of wire designed by Harry Bertoia, circa 1952. From the "Industrial Design" exhibit at the Grand Palais, Paris. Courtesy The Musee des Arts Decoratif.

Conoid cushion chair designed by George Nakashima, 1962. Courtesy Bill Holland.

High stool from the "Leonardo collection" designed by Paul Tuttle, circa 1979. From the "Industrial Design" exhibit at the Grand Palais, Paris. Courtesy The Musee des Arts Decoratif.

Conoid table and chair designed by George Nakashima, 1962. Courtesy Bill Holland.

Stacking stool designed by Florence Knoll, circa 1950. From the "Industrial Design" exhibit at the Grand Palais, Paris. Courtesy The Musee des Arts Decoratif.

Folding screen (model FSW) designed by Charles and Ray Eames for Herman Miller, Inc., circa 1946. From the "Industrial Design" exhibit at the Grand Palais, Paris. Courtesy The Musee des Arts Decoratif.

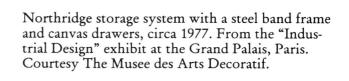

Northridge storage system with a steel band frame and canvas drawers, circa 1977. From the "Industrial Design" exhibit at the Grand Palais, Paris. Courtesy The Musee des Arts Decoratif.

Plant stand made of a tree branch, knots, and roots, circa 1990. From the "Industrial Design" exhibit at the Grand Palais, Paris. Courtesy The Musee des Arts Decoratif.

Combination chairs designed by Gilbert Rohde, circa 1940. From the "Industrial Design" exhibit at the Grand Palais, Paris. Courtesy The Musee des Arts Decoratif.

Settee designed by Alexander Gerard for Herman Miller. Courtesy Don Treadway Gallery.

Jewelry chest designed by George Nelson for Herman Miller.--Rocking Shell armchair designed by Charles and Ray Eames.--#400 Storage Unit designed by Charles and Ray Eames. Courtesy Don Treadway Gallery.

Hall console designed by Gio Ponti. Courtesy Don Treadway Gallery.

Table and chair designed by Frank Lloyd Wright for the Price Tower. Courtesy Don Treadway Gallery.

Telephone stand designed by Alexander Girard for Herman Miller. Courtesy Don Treadway Gallery.

Desk designed by George
Nelson for Herman Miller.
Courtest Don Treadway
Gallery.

Storage unit designed by
Charles Eames for Herman
Miller Furniture Com-
pany. Courtesy Don
Treadway Gallery.

Child's chair and stool designed by Charles and Ray Eames for Evans Product.--Storage unit designed by Charles Eames for Herman Miller.--House of Cards designed by Charles Eames for Ravensburger Spiele. Courtesy Don Treadway Gallery.

Aluminum Group chair and ottoman designed by Charles Eames and introduced in 1958. The design began on the back of an envelope as a sketch of a connection between metal and fabric. That connection, between the structural side ribs of die-cast aluminum and the two layers of heat-sealed fabric that form the seat pad, determined the structure of the chair. Aluminum group chairs were available in low- or high-back versions and a lounge that includes an attached headrest.

Structurally identical to the Aluminum Group is the Soft Pad Group that was developed in 1969. It has one large soft pad on the seat and two, three, or four smaller soft pads to cushion the back. Courtesy Herman Miller Archives, Zeeland, Michigan.

Lounge chair designed by George Nakashima, 1962. Courtesy Bill Holland.

Bench with a back designed by George Nakashima, 1958. Courtesy Bill Holland.

Conoid bench with a back designed by George Nakashima, 1962. Courtesy Bill Holland.

Night stand and end table designed by Gilbert Rohde. Courtesy Savoia's Auction Incorporated.

Coffee table designed by George Nelson for Herman Miller. Courtesy Savoia's Auction Incorporated.

Segmented base tables designed by Charles Eames. Courtesy Herman Miller, Incorporated, Zeeland, Michigan.

Universal base tables designed by Charles Eames.
Courtesy Herman Miller, Incorporated, Zeeland,
Michigan.

Contract base tables designed by Charles Eames.
Courtesy Herman Miller, Incorporated, Zeeland,
Michigan.

Lafonda base table designed by Charles Eames.
Courtesy Herman Miller, Incorporated,
Zeeland, Michigan.

Wire base table designed by Charles Eames.
Courtesy Herman Miller, Incorporated,
Zeeland, Michigan.

Walnut stool designed by Charles Eames.
Courtesy Herman Miller, Incorporated,
Zeeland, Michigan.

Free-form cork top coffee table designed by Paul Frankl. Courtesy Savoia's Auction Incorporated.

Dressing table, part of a nine-piece bedroom set designed by George Nelson. Courtesy Savoia's Auction Incorporated.

Glass topped coffee table designed by Noguchi for Herman Miller, Inc. Photograph courtesy of Herman Miller, Inc.

Dressing vanity #3920 designed by Gilbert Rohde, circa 1935, for the Herman Miller Furniture Company. The case of ebonized plywood and rosewood veneers is embellished with brass, Plexiglas, Fabricoid, and glass elements.

Storage units by Charles Eames, 1950. Courtesy Herman Miller Incorporated Archives, Zeeland, Michigan.

Bibliography

Aslin, E. *The Aesthetic Movement: Prelude to Art Nouveau*. New York: 1969.

Bavaro, J. J. and T. L. Mossman. *The Furniture of Gustav Stickley*. New York: Van Nostrand, Reinhold, 1982.

Caplan, R. *The Design of Herman Miller*. New York: Whitney, 1976.

Cathers, D. M. *Furniture of the American Arts and Crafts Movement*. New York: New American Library, 1981.

Comstock, Helen. *American Furniture 17th, 18th, and 19th Century Styles*. New York: Viking Press, 1972.

Dal Farbo, M. *Furniture for Modern Interiors*. New York: Reinhold, 1954.

Del Farbo, M. *Modern Furniture: Its Design and Construction*. New York: Reinhold, 1958.

Darling, S. *Chicago Furniture: Art, Craft and Industry 1833-1983*. Domergur, D. *Artist Design Furniture*. New York: Harry N. Abrams, 1985.

Drexler, A. *Charles Eames: Furniture From The Design Collection*. New York: Musuem of Modern Art, 1973.

Dubrow, Richard and Eileen. *American Furniture of the 19th Century*. Exton: Schiffer Publishing Ltd., 1983.

Duncan, Alistair. *American Art Deco*. New York: Harry N. Abrams, 1968.

Duncan, Alistair. *Art Nouveau Furniture*. New York: Clarkson N. Potter, Inc., 1982.

Fairbanks, Jonathan and Elizabeth B. Bates. *American Furniture 1620 to the Present*.

Gems of the Centennial Exhibition. New York: D. Appleton & Company, 1877.

Grief, M. *Depression Modern: The Thirties Style in America*. New York: Universe, 1975.

Hanks, David. *Innovative Furniture in America from 1880 to the Present*. New York: Horizon Press, 1981.

Howe, K. and David Warren. *The Gothic Revival Style in America 1830-1870*. Houston: The Museum of Fine Arts, 1976.

In Pursuit of Beauty. New York: The Metropolitan Museum of Art and Rizzoli, 1986.

Johnson, D. *American Art Nouveau*. New York: Harry N. Abrams, 1979.

Johnson, M. et al. *19th Century America*. New York: Metropolitan Museum of Art, 1970.

Lynes, R. *The Taste Makers: The Shaping of American Taste*. New York: Dover Puvlications, 1949.

Macleod, R. *Charles Rennie Mackintosh*. Feltham, 1968.

Madigan, Mary Jane. *19th Century Furniture*. New York: Roundtable Press, 1982.

Maker, J. T. *The Twilight of Splendor*. New York: Little Brown & Co., 1975.

Makinson, R. L. *Greene and Greene, Furniture and Related Designs*. New York: Peregrine Smith , Inc., 1979.

Mayhew, E. and M. Meyers. *A Documentary History of American Interiors*. New York: Charles Scribner, 1980.

Moody, E. *Modern Furniture*. New York: E. P. Dutton, 1966.

Naylor, G. *The Bauhaus*. New York: E. P. Dutton and Vista, 1974.

Otto, C. *American Furniture of the 19th Century*. New York: Viking, 1965.

Page, M. *Furniture Designed by Architects*. New York: Whitney Museum, 1980.

Sparke, P. *Furniture- Twentieth Century Design*. London: Bell & Hyman, Ltd.

Spencer, R. *The Aesthetic Movement*. New York, 1972.

Three Centuries of American Art. Philadelphia: Philadelphia Museum of Art, 1976.

Weber, E. *Art Deco in America*. New York: Bison, 1985.

Wilk, C. *Marcel Bruer: Furniture and Interiors*. New York: Museum of Modern Art, 1980.

Wilson, R., D. Pilgrim and R. Murray. *The American Renaissance 1876-1917*. New York: The Brooklyn Museum, 1979.

Values Reference

Pg.	Pos.	Value	Pg.	Pos.	Value	Pg.	Pos.	Value	Pg.	Pos.	Value
5	BL	$1,200-1,500	43	TL	$800-1,200		TR	$3,000-5,000		BR	$10,000-15,000
	BC	$400-600		TR	$1,200-1,500 ea.		BL	$6,000-8,000	73	BR	$8,000-12,000 suite
	BR	$1,500-2,000		CL	$3,000-5,000		BR	$3,000-5,000	74	TR	$6,000-8,000
6		$2,000-2,500		BL	$1,000-1,500	58	TR	$3,000-5,000 ea.		BL	$6,000-8,000
9	TL	$200-300		BC	$1,500-2,000		BR	$5,000-7,000 ea.	75	TL	$4,000-6,000
	BR	$2,000-2,500		BR	$1,000-1,500	59	TL	$3,000-5,000		C	$2,000-3,000
10	TL	$300-500	44	TL	$600-800		TR	$2,000-3,000		BL	$2,000-3,000 ea.
	BR	$100-200		BR	$300-400		CL	$4,000-6,000	76	TL	$2,000-2,500
32	BR	$5,000-7,000	45	T	$400-600		BR	$1,500-2,000		TR	$3,000-5,000
33	TL	$1,500-2,000		B	$2,000-2,500	60	TL	$4,000-6,000		BL	$1,000-1,500
	TR	$2,000-2,500	46	TR	$800-1,200		TR	$6,000-8,000		BC	$2,000-3,000
	CR	$300-500		B	$3,000-5,000 ea.	61	TL	$2,000-3,000		BR	$800-1,200
	B	$5,000-7,000 suite	47	TL	$4,000-6,000		TR	$3,000-5,000	77	TL	$3,000-4,000
34	TR	$2,000-2,500		TR	$3,000-5,000		BL	$4,000-6,000		TR	$2,000-3,000
	BL	$3,000-5,000		BL	$5,000-7,000		BR	$2,000-3,000		BL	$3,000-5,000
35	TL	$600-800	48	TL	$10,000-15,000	62	TR	$2,000-3,000		BC	$3,000-5,000
	BL	$1,200-1,500	49	TL	$1,500-2,000		BL	$10,000-15,000+	78	TL	$3,000-5,000
36	TL	$700-900		CR	$6,000-8,000+	63	TR	$3,000-5,000		BR	$12,000-15,000+
	TR	$300-500	50	TL	$4,000-6,000		BL	$4,000-6,000	79	TL	$5,000-7,000
	B	$1,500-2,000		TR	$4,000-6,000		BR	$1,000-1,500		TR	$4,000-6,000
37	T	$4,000-6,000		BL	$3,000-5,000	64	TL	$3,000-5,000		BL	$3,000-5,000
	BL	$2,000-3,000		BR	$3,000-5,000		C	$6,000-8,000		BR	$10,000-15,000
	BR	$1,500-2,000	51	TR	$10,000-15,000+		BR	$5,000-8,000	80	TL	$5,000-7,000+
38	CL	$100-200		CL	$2,000-3,000	65	TL	$2,500-3,500		TR	$6,000-8,000+
	TR	$300-500		BR	$3,000-5,000		TR	$1,500-2,000		CL	$5,000-7,000+
	BR	$300-500	52	TL	$2,000-3,000		BL	$10,000-15,000+		BR	$10,000-15,000+
39	T	$10,000+		BR	$5,000-7,000		CR	$600-800	81	TL	$3,000-5,000
	BL	$600-800	53	TR	$1,500-2,000		BR	$1,500-2,000		TR	$4,000-6,000
	CR	$2,000-2,500 ea.		CL	$800-1,200 ea.	66	TL	$6,000-8,000	82	TL	$3,000-5,000
	BR	$2,000-2,500 ea.		B	$4,000-6,000+		TR	$600-800		BL	$3,000-5,000
40	TL	$4,000-6,000	54	BL	$2,000-3,000		CL	$1,500-2,000		BR	$5,000-7,000+
41	TL	$1,000-1,500		BR	$4,000-6,000		CR	$6,000-8,000+	83	TL	$6,000-8,000+
	BR	$600-800	55	TL	$10,000-15,000+ ea.	67	TR	$3,000-5,000		BR	$3,000-5,000+
42	TL	$1,500-2,000		B	$3,000-5,000		CL	$10,000-15,000+	84	CL	$3,000-5,000
	TC	$2,000-3,000	56	TL	$2,000-3,000	68		$10,000-15,000+		BR	$3,000-5,000
	TR	$1,500-2,000		TR	$10,000-15,000+	69		$20,000-30,000+	85	TL	$2,000-3,000
	CL	$8,000-12,000		BL	$5,000-7,000	70		$15,000-20,000+		TR	$10,000-15,000
	BR	$300-500	57	TL	$3,000-5,000+	71		$20,000-25,000+		B	$10,000-15,000+
	BL	$200-300				72	TL	$8,000-10,000+	86	TL	$3,000-5,000+

Pg.	Pos.	Value	Pg.	Pos.	Value	Pg.	Pos.	Value	Pg.	Pos.	Value
	TR	$4,000-6,000+	103	CL	$10,000-15,000+		BR	$3,000-5,000		TR	$600-800
	BL	$8,000-12,000+		BL	$10,000-15,000+	121		$15,000-20,000 set		BL	$800-1,200
87	TL	$15,000-20,000		BR	$6,000-8,000+	122	TR	$30,000-40,000+		BR	$300-500
	CR	$12,000-15,000+	104	TL	$1,000-1,500		BR	$1,500-2,000	138	TL	$1,500-2,000
	BL	$12,000-15,000		TR	$3,000-5,000	123	T	Rare			$800-1,200
88	TL	$10,000-15,000+		BL	$2,000-3,000	124	TL	$10,000-15,000		TR	$1,200-1,500 chair
	TR	$4,000-6,000		BR	$2,000-3,000		TR	$5,000-7,000			$1,500-2,000 table
	BL	$4,000-6,000+	105	TL	$2,000-3,000		TR	$2,000-3,000 chair		BL	$1,500-2,000
	BR	$3,000-5,000+		TR	$4,000-6,000			$1,500-2,000 plant		BR	$2,500-3,500
89	TL	$10,000-15,000+		C	$3,000-5,000			stand	139	TL	$5,000-7,000
	CR	$10,000-15,000+		BL	$1,000-1,500 ea.			$6,000-8,000 table		CR	$1,200-1,500
	BL	$7,000-9,000+		BR	$600-800		BL	$15,000-20,000		BL	$2,000-3,000
90	TL	$1,500-2,000	106	TL	$2,000-3,000		BR	$10,000-15,000	140	TR	$3,000-5,000
	TR	$3,000-5,000		TR	$4,000-6,000	125	T	$15,000-20,000 set		CL	$4,000-6,000
	BL	$2,000-3,000		BL	$4,000-6,000		B	$4,000-6,000		BR	$2,000-3,000
	BR	$3,000-5,000		BR	$4,000-6,000	126	T	$10,000-15,000	141	TR	$4,000-6,000
91	T	$400-600 each	107	TL	Rare		C	$4,000-6,000		CL	$3,000-5,000
	BL	$2,000-3,000		CR	Rare		B	$4,000-6,000		BR	$2,000-3,000
	BC	$4,000-6,000		BL	$20,000-30,000+	127	TL	$4,000-6,000	142	TR	$800-1,200
	BR	$600-800		BR	$10,000-15,000+		TR	$1,500-2,000 table		CL	$1,500-2,000
92	TL	$200-300	108	TL	$15,000-20,000+			$8,000-10,000		BR	$400-600
	BL	$4,000-6,000		TR	$15,000-20,000+			chest	143	T	$1,500-2,000
	CR	$3,000-5,000		BL	$20,000-30,000+		BL	$10,000-15,000		BL	$400-600 each
93	TR	$200-300		BR	$2,000-3,000		BR	$2,500-3,500		BR	$800-1,200
	C	$200-400	109	CR	$20,000-30,000+	128	TL	$500-700	144	T	$5,000-7,000
	B	$1,500-2,000		BL	$10,000-15,000+		TR	$800-1,200		BL	$3,000-5,000
94	TL	$1,000-1,500	110	TL	$10,000-15,000		BL	$200-300		BR	$3,000-5,000
	TR	$800-1,200		TR	$10,000-15,000		CR	$200-300	145	TL	Rare
	BL	$1,500-2,000		BL	$20,000-30,000+	129	TL	$10,000-15,000		TR	$1,500-2,000
	BR	$1,500-2,000		BR	$20,000-30,000+		TR	$1,500-2,000		BL	$2,000-3,000
95	TL	$700-900	111	TL	$4,000-6,000+		B	$6,000-8,000		BR	$2,000-3,000 ea.
	BL	$2,000-3,000		TR	$6,000-8,000	130	T	$15,000-20,000	146	TL	$1,500-2,000
	BR	$1,200-1,500		BL	$4,000-6,000		C	$10,000-15,000		TR	$3,000-5,000
96	TL	$3,000-5,000	112	TL	$20,000-25,000+		B	$10,000-15,000		BL	$5,000-7,000
	TR	$3,000-5,000		TR	$3,000-5,000	131	TL	$8,000-12,000	147	TL	$2,000-3,000 ea.
	BL	$1,000-1,500		BR	$10,000-15,000		BL	$6,000-8,000		TR	$3,000-5,000
	BR	$600-800	113	T	$2,000-3,000		R	$3,000-5,000			$1,200-1,500
97	TL	$2,000-3,000		CL	$1,500-2,000	132	T	$200-300 chair		B	$4,000-6,000
	TR	$1,000-1,500		CR	$800-1,000			$700-900 bench	149	T	Rare
	BL	$600-800		BL	$600-800		CL	$200-300		CL	Rare
	BR	$1,200-1,500		BR	$500-700		CR	$75-150		BR	$6,000-8,000
98	TL	$100-200	114	T	Rare		BL	$300-500	150	TL	$2,000-3,000
	TR	$800-1,200	116	T	$600-800		BR	$75-150		TR	$4,000-6,000
	BL	$200-300		B	$600-800	133	C	$6,000-8,000		BR	$4,000-6,000
	BR	$600-800	117	TL	$600-800		B	$400-600	151	TL	$10,000-15,000
99	TL	$800-1,200		TR	$1,200-1,500	134	TR	$4,000-6,000		TR	$4,000-6,000
	TR	$400-600		BL	$300-500		CL	$2,000-2,500		BL	$10,000-15,000
	BL	$800-1,200		BR	$800-1,200		BR	$4,000-6,000		BR	$8,000-12,000
	BR	$800-1,200	118	L	$800-1,200	135	TR	$1,500-2,000	152	TR	$5,000-7,000
100	TL	$500-700		R	$2,000-3,000		CL	$2,000-2,500		BL	$4,000-6,000
	TR	$1,000-1,500	119		$15,000-20,000+		BR	$300-500		BR	Rare
	BL	$2,500-3,000	120	TL	$700-900	136	TR	$200-300	153	TL	Rare
	BR	$1,000-1,500 set		TR	$700-900		BR	$1,500-2,000		TR	$15,000-20,000
101		$1,000-1,500		BL	$700-900	137	TL	$800-1,200		BR	$2,000-3,000

Pg.	Pos.	Value	Pg.	Pos.	Value	Pg.	Pos.	Value	Pg.	Pos.	Value
154	TL	$3,000-5,000		TR	$600-800		TR	$600-800		CL	$200-300
	TR	$2,000-3,000		B	$2,000-3,000		BR	$1,500-2,000		BR	$2,000-3,000
	B	$4,000-6,000	162	TR	$200-300	178	TL	$700-900	187	TL	$500-700
155	T	$8,000-12,000		BL	$400-600		TR	$1,200-1,500		R	$1,000-1,500
	BL	$2,000-3,000	163	TR	$2,000-3,000		B	$2,000-3,000	188	T	$3,000-5,000
	BR	$6,000-8,000		B	$6,000-8,000 set	179	TL	$5,000-7,000 chest		B	$2,000-3,000
156	TL	$8,000-12,000	164	TL	$1,500-2,000			$2,000-2,500 night	189		$1,000-1,500
	TR	$2,000-3,000		BL	$400-600			stand			jewelry chest
	BL	$500-700		R	$10,000-15,000		B	$1,000-1,500 chair			$4,000-6,000
	BR	$10,000-15,000	165	TL	$3,000-5,000			$1,500-2,000			storage unit
157	TL	$3,000-5,000		BL	$6,000-8,000			cabinet			$600-800 chair
	TR	$7,000-9,000 chair		R	$4,000-6,000	180	T	$1,500-2,000	190	T	$2,000-3,000
		$1,200-1,500 table	166	B	$500-700			loveseat		BL	$400-600
	BL	$6,000-8,000 chair	167	T	$2,000-3,000			$300-500 table		BR	$8,000-10,000
		$1,200-1,500 foot		BL	$800-1,200			$600-800 chair	191	T	$1,000-1,500
		stool		BR	$800-1,200		B	$1,000-1,500 stool		BL	$4,000-6,000
		$8,000-12,000	168	TL	$2,000-3,000			$1,000-1,500 table	192	$	$400-600 child's
		table		TR	$1,500-2,000			$2,000-3,000 chair			chair
158	TL	$3,000-5,000		B	$4,000-6,000	181	TL	$15,000-20,000		$	$200-300 stool
	TR	$4,000-6,000 lg.	169	T	$5,000-7,000			chair		$	$8,000-10,000
		chair		C	$5,000-7,000			$8,000-12,000			storage unit
		$1,200-1,500 table		B	$4,000-6,000			skylight	193	T	$600-800
		$500-700 sm. chair	170	TL	$3,000-5,000		CR	$100-150		B	$2,500-3,500
	B	$2,500-3,500 table		TR	$3,000-5,000		BL	$150-200 chair	194	T	$3,000-5,000
		$10,000-15,000		BR	$2,000-2,500			$2,000-3,000		B	$7,000-9,000
		sofa	171	TL	$10,000-15,000			storage unit	195	TL	$300-500 night
159	T	$15,000-20,000		TR	$12,000-15,000	182	T	$1,500-2,000			stand
	CL	$10,000-15,000		B	$10,000-15,000	183	TL	$500-700 chair			$400-600 table
	CR	$4,000-6,000	172	B	$1,500-2,000 ea.			$2,000-3,000		TR	$800-1,200
	B	$1,200-1,500	173	B	$700-900 each			screen		B	$600-3,000 ea.
		umbrella stand	174	T	$8,000-10,000			$400-600 table	196	T	$200-300 ea.
		$1,000-1,500		BL	$4,000-6,000		CR	$1,500-2,000		B	$200-300 ea.
		wastebasket		BR	$2,000-3,000		BL	$800-1,200	197	TL	$400-600
		$1,200-1,500	175	T	$2,500-3,500	184	TL	$50-75		CR	$600-800
		mirror		BL	$2,500-3,500		TR	$10,000-15,000		B	$600-800 ea.
160	TL	$1,500-2,000		BR	$2,000-3,000 ea.		B	$50-75 ea.	198	TL	$1,200-1,500
	TR	$4,000-6,000	176	T	$3,000-5,000	185	T	$300-500		C	$1,500-2,000
	BL	$300-500		CR	$1,200-1,500		BL	$600-800		B	$2,000-3,000
	BR	$500-700		BL	$100-200		BR	$2,000-3,000	199		$5,000-7,000+
161	TL	$400-600	177	TL	$800-1,200	186	T	$6,000-8,000	200		$2,000-10,000

Index